# Management for Professionals

The Springer series *Management for Professionals* comprises high-level business and management books for executives. The authors are experienced business professionals and renowned professors who combine scientific background, best practice, and entrepreneurial vision to provide powerful insights into how to achieve business excellence.

Stefan Aichbauer • Martina Buchhauser •
Agnes Erben • Sven Steinert •
Detlef Tietze • Emilia Wiking

# Responsible Procurement

## Leading the Way to a Sustainable Tomorrow

Springer

Stefan Aichbauer
H&Z Unternehmensberatung AG
Munich, Germany

Martina Buchhauser
H&Z Unternehmensberatung AG
Munich, Germany

Agnes Erben
H&Z Unternehmensberatung AG
Munich, Bayern, Germany

Sven Steinert
H&Z Unternehmensberatung AG
Munich, Germany

Detlef Tietze
H&Z Unternehmensberatung AG
Munich, Germany

Emilia Wiking
H&Z Unternehmensberatung AG
Munich, Germany

ISSN 2192-8096 ISSN 2192-810X (electronic)
Management for Professionals
ISBN 978-3-030-98639-1 ISBN 978-3-030-98640-7 (eBook)
https://doi.org/10.1007/978-3-030-98640-7



This Springer imprint is published by the registered company Springer Nature Switzerland AG
The registered company address is: Gewerbestrasse 11, 6330 Cham, Switzerland

*For all procurement professionals that strive to be at the cutting edge of supply chain sustainability.*

# Foreword

If Formula 1, perhaps the most excessive of sports, can clean up its act, surely the rest of us can as well.

**Nico Rosberg**

To be honest, sustainability was not one of the things I used to think about when I was strapped into the cockpit of my Formula One car and driving at speeds in excess of 300 km/h around iconic racetracks. I had other more immediate and life-changing thoughts to deal with then! But since I stepped away from the sport, it dominates my thoughts. As an entrepreneur and investor in green technology and alternative mobility, sustainability, doing better, and positive change are the three pillars that guide all my decisions.

We're seeing the spread of similar values across society. Seventy percent of young consumers will choose a rival product over their favorite brand if the rival is more value driven and sustainable. This is exactly the kind of attitude shift that we need, but it has to go further. We need to see it as part of business culture. In fact, I recently saw for myself how the kids of some of the world's biggest tech CEOs have managed to push this through by saying "Dad, what you're doing at XYZ company is unacceptable. You have the power so make the change—you have to do it!" I also believe social media can play a big role here. It can help us to keep an eye on what's happening to all companies in the supply chain, expose any shortcomings, and keep the pressure on everyone to do their part.

There is a real urgency attached to this in more ways than one. The latest *Intergovernmental Panel on Climate Change* (IPCC) report is alarming because it says we could reach the 1.5–2 °C rise in global average temperatures during this decade. We in developed countries need to do more to prevent this and we can and should lead the way. We have the potential for the greatest world penetrating innovation. And in fact, it will help us to stay ahead economically because those nations and companies who trailblaze net-zero tech will futureproof themselves against an uncertain future.

It's a huge challenge and it'll only work if we work together. I know this as a race driver. Alone nothing was possible—there were 1500 colleagues in my team for just two race cars. Everyone's role was crucial in winning the big prize. We complemented each other and worked together to be the best version of ourselves.

In my championship winning year, I even had to let my teammate and greatest rival overtake me during the Monaco Grand Prix as he was the only one who had a chance to win the race for the team. Society and business can do the same—we have to keep an open mind and collaborate at all levels, even with our rivals, when necessary, if it's the smart way to long-term success.

I see this as the biggest, most important, and most challenging race I've been involved in—the race to zero. If we're successful, we'll all be winners. But if we fail, nobody knows exactly what the consequences will be, but looking at recent extreme weather events in Europe and around the world gives us some idea. That's not a future I want.

We're in this together, and we can all contribute to a sustainable future. After all, contribution is a basic human need, and it can be very fulfilling. It's what drives me today. It's also very motivating to think that I inspire my kids because it's their future and the future of all kids that I'm investing in. This is why I love backing companies that are purpose-driven, companies who really care. It's now clear that as a business you can make a substantial contribution to sustainability at the same time as making a profit—you don't have to choose one or the other.

Ironically, the need for a sustainable future is one of the driving forces behind Formula 1, which has committed to developing sustainable fuels, becoming net zero by 2030, and to produce zero waste on and off track. If Formula 1, perhaps the most excessive of sports, can clean up its act, surely the rest of us can as well.

**Nico Rosberg** is known as the Formula 1 World Champion of 2016. Since ending his active driving career, he has been passionate about sustainability. As an entrepreneur and investor in green technologies and alternative mobility start-ups, he stands for visionary innovations that make the world a little better. He is also involved in numerous projects around the world promoting environmental protection, health, equality, and justice. In 2019, he founded the GREENTECH FESTIVAL, a global platform for pioneering sustainable ideas, which takes place annually in Berlin, and—as of 2021—New York and London. Using his former sport as an impactful vehicle, Nico Rosberg has founded Rosberg X Racing, a team that is driven to inspire positive change. The team competes in Extreme E, a racing series that aims to draw attention to the risks and consequences of global climate change. Rosberg has received numerous awards for his entrepreneurial commitment. He lives in Monaco with his wife and two daughters.

Monaco, Monaco Nico Rosberg

# Preface

Sustainability does provide the ideal platform to redefine procurement's role as the orchestrator of the supplier ecosystem and that's why sustainable business growth is what procurement should be helping the business to deliver.

**Mark Perera, Chief Executive Officer, Vizibl**

It has taken a long time to get to where we are today, but there is now an unstoppable force, created by a collective of diverse stakeholders, driving the sustainability agenda. Consumers want green products made by workers who are treated with fairness, dignity, and respect no matter where they are in the world; institutional investors are increasingly selective about where they allocate their vast resources; and sustainability is being embraced in boardrooms across the globe.

At the time of writing, more than 2,000 businesses and financial institutions were working with the Science Based Targets initiative (SBTi) to reduce their emissions in line with climate science (SBT, 2021). Leaders from some of the biggest companies among this powerful coalition were present at the 2021 United Nations Climate Change Conference (Cop 26) in Glasgow, UK, "with credible evidence of their progress (*toward a carbon neutral economy*) as well as a sincere and urgent desire to push governments to go further and faster" (Mendiluce, 2021).

We believe this is exactly how it should be. As consultants and leaders (and people who care about the world around them) with a wealth of experience of business transformation, procurement, and sustainability, it is our view that business leaders should not wait until legislation or public opinion forces them reluctantly to make sustainability a core part of their corporate strategies. It is our belief that sustainability is as much an opportunity to build better businesses as it is about taking action on climate change and looking after the welfare of all workers along the supply chain.

We also believe that more could and should be done, which is one purpose for writing this book.

## Procurement: The Front line of Corporate Sustainability

Above all, we strongly believe that it is procurement that should be at the vanguard—*a group of people leading the way in new developments or ideas*—of business sustainability.

Throughout Europe many companies have started to get their internal operations under control by mitigating their risks, building sustainability into working procedures, and strictly following health and safety regulations. However, it is externally, along the supply chain to the point where raw materials are extracted, where the real challenges lie. It is no longer acceptable that procuring goods comes at the expense of damage to the environment or to the people who live where extraction takes place. A company's sustainability standards can only truly be as good as those along its entire supply chain.

The primary purpose of this book is to encourage procurement professionals to combine bold moves with common sense and go beyond basing supply chain decisions on price alone. There is no simple remedy with which to solve this highly complex situation. Decision makers and all other stakeholders need to put into effect a variety of actions that will gradually make sustainability as important as any other essential function.

## A Sustainable Business Starts from the Top

If procurement is at the front line of corporate sustainability, the people implementing the strategy on a day-to-day basis need to know that those at the top of the company are right behind them. Leadership plays a pivotal role in a sustainability transformation. This is as much about motivation as it is about seeing the big picture and the opportunities that can be found in running a sustainable business. There is a section of the book dedicated to this vital leadership role and how to inspire from the front.

### Sustainability in Action

> Now it's about what we're doing to humanity in the world to make it a better place. Now we see CEOs at business roundtables around the world talking about the wider stakeholders, and that's not just coming from a sustainability environmental party, about employees, about our customers and it's about our suppliers and that there has to be purpose. Purpose needs to be authentic to a company and in a brand. You just can't copy and paste it from another organization. Purpose . . . that's what gets the hearts and minds of employees or customers or shareholders as well, which I look at as an investment. Ultimately if you have good ESG scoring around that, that also means you shouldn't have bad press coming out about your organization for doing slave labor or ruining the planet or no treating people with disrespect which ultimately would drag your share price down.
>
> **Mark Perera, Chief Executive Officer, Vizibl**
>
> Vizibl is a digital procurement platform that allows global organizations to unlock the true value of their supplier relationships.

### Sustainability in Action

I'm a huge fan of technology ... but also, oftentimes, it is behavioral change and mindset change that need to occur within the companies. You need to get and win the hearts and minds of the management team, which sometimes may need to be accelerated by the application of tough economic pressure, in order to instill that 'doing-it-right' culture in their companies. Audits yield interesting feedback in closing meetings when discussing findings and if you then have an OEM on the other side that says, 'Mrs. Battery Supplier, your performance has been completely unacceptable in this audit, let's figure this out and let's get you guys to change your way of looking at this issue and your approach to doing business.' It's that sort of qualitative engagement that often actually has the biggest impact because once you get management's attention to a key issue, once it becomes a boardroom issue and once it's linked to their contracts and their procurement initiatives, suddenly, you have really strong levers and that's how you change the world.

**Dr Nicholas Garrett, Co-founder and Chief Executive Officer, RCS Global Group**

RCS Global Group is a proven leader in data driven ESG performance, responsible sourcing, and responsible mining assurance.

## Who This Book Is for

This book is aimed in particular at those working in procurement. They manage significant proportions of the money their companies spend and thus have the power (and the responsibility) to make informed supply chain decisions that satisfy the triple bottom line, the measurement of social and environmental impacts alongside profits. The book contains advice and guidance, as well as real world examples, to help procurement practitioners do the right thing. After all, they are at the forefront of change for a sustainable corporate future.

But it is not just aimed at procurement. The purpose of this book is to inspire decision-makers, managers, and all company representatives who have roles in procurement and the supply chain who want to contribute to a sustainable future for the companies they work for, and for themselves. Whether the reason is to satisfy investor or consumer demands, adhere to regulations, appease social pressures, become a "greener" and more resource-efficient organization, or simply because the motivation is intrinsic, the end results will be the same.

## You Are Not Alone

You are not alone when it comes to developing sustainable business strategies. Sustainability is regarded as the new frontier in business so there are numerous ways of getting help. There are dedicated organizations whose raison d'être is to guide businesses through their sustainability options. There are structured methodologies, and, not surprisingly, technology has started to play a major role. We will highlight the latest and greatest of these, show real-life case studies, and offer contacts and practical advice to kickstart the sustainability journey.

With mounting pressure coming from legislators, consumers, and investors, it should come as no surprise that 90% of executives think sustainability is important yet only 60% of companies have a sustainability strategy (Unruh et al., 2016). Often, companies that speak of being sustainable are lacking when it comes to implementation. This needs to change.

The biggest risk of all is not taking action.

## Bibliography

Mendiluce, M. (2021). *The street and the boardroom are closer than they have ever been on climate*. Accessed November, 22, 2021, from Climate Home News: https://climatechangenews.com/2021/11/19/street-boardroom-closer-ever-climate/

SBT. (2021). *Companies taking action, science based targets*. Accessed November, 22, 2021, from https://sciencebasedtargets.org/companies-taking-action

Unruh, G., KIiron, D., Kruschwitz, N., Reeves, M., Rubel, H., & Meyer Zum Felde, A. (2016). Investing for a sustainable future. *MIT Sloan Management Review*. Accessed July, 27, 2021, from https://sloanreview.mit.edu/projects/investing-for-a-sustainable-future/

Munich, Germany

Stefan Aichbauer
Martina Buchhauser
Agnes Erben
Sven Steinert
Detlef Tietze
Emilia Wiking

# Acknowledgements

There is no denying that writing a book on sustainability is a challenging undertaking due to its continually evolving nature, especially post COP26. With that in mind, we would like to thank our various co-authors and contributors for making the process much smoother:

Most importantly, we want to thank our supporters in the industries who have helped in making our book lively and demonstrating sustainability in action. Thank you all for taking the time.

**Simon Anholt**, Chair of the Steering Group, Inner Development Goals Initiative

Inner Development Goals (IDGs) is an initiative that provides a blueprint of the capabilities, qualities, and skills needed to achieve the 17 SDGs.

**Marielle Beyer**, Head of Global Procurement, Roche

With its purpose of "Doing now what patients need next", Roche has a 120-year history of advancing the field of medicine and bringing novel treatments and diagnostics to patients.

**Peter Carlsson**, Co-founder and Chief Executive Officer, Northvolt

Northvolt is striving to produce the greenest battery in the world with a minimal carbon footprint and the highest ambitions for recycling to enable the European transition to renewable energy.

**Nils Eichberger**, Vice President of Procurement, LBBW (Landesbank Baden-Württemberg)

LBBW is a Mittelstand-minded universal bank with deep roots in Baden-Württemberg and a partner for medium-sized companies, institutional customers, savings banks, and retail customers.

**Dr Nicholas Garrett**, Co-founder and Chief Executive Officer, RCS Global Group

RCS Global Group is a proven leader in data driven ESG performance, responsible sourcing, and responsible mining assurance.

**Sakuto Goda**, Principal, Nomura Research Institute

Nomura Research Institute (NRI) is a leading consulting firm and systems integrator from Japan with expertise in the formulation of strategies, globalization of business operations and planning.

**Nina Göntgen-Voss**, Director of Sustainability, Forto

Forto delivers a highly transparent, frictionless, and sustainable digital supply chain. Their platform addresses the entire process stream, supporting customers with greater visibility, insight, and control.

**Fredrik Gustafsson**, Senior Manager of Sustainability Transformation, Cordial

Cordial is a management consulting firm based in Stockholm, Sweden. Its main focus is strategy, transformation, and business development with sustainability as one of the key areas.

**Oda Heister**, Managing Director, Ashoka Germany

Ashoka identifies and supports the world's leading social entrepreneurs, learns from their innovations, and mobilizes a global network to build an "everyone a changemaker world."

**Jan Artem Henriksson**, Chief Executive Officer, Inner Development Goals Initiative

Inner Development Goals (IDGs) is an initiative that provides a blueprint of the capabilities, qualities, and skills needed to achieve the 17 SDGs.

**Thomas Janvier**, Vice President of Procurement, INNIO

INNIO is a leading provider of renewable gas, natural gas, and hydrogen-based solutions and services for power generation and gas compression at or near the point of use.

**Henrik Larsen**, Chief Procurement Officer, A. P. Møller-Mærsk

A. P. Møller-Mærsk is an integrated container logistics company and member of the A.P. Møller Group. With a dedicated team of over 80,000 people, they enable global trade for a growing world.

**Chaojun Li**, Global Head of Sustainability, WS Audiology

WS Audiology develops, manufactures, sells, and distributes hearing aids. WS Audiology's ambition as a global leader is to unlock human potential by making wonderful sound part of everyone's life.

**Bastian Litterscheid, Managing Director, Berger Fahrzeugtechnik**

BERGERecotrail builds the most stable and lightest semi-trailers on the market and continues to pioneer for sustainably successful logistics operations.

**Dr María Mendiluce**, Chief Executive Officer, We Mean Business Coalition

We Mean Business Coalition is a global non-profit coalition working with the world's most influential businesses to take action on climate change.

**Mark Perera**, Chief Executive Officer, Vizibl

Vizibl is a digital procurement platform that allows global organizations to unlock the true value of their supplier relationships.

**Markus Strangmüller**, Corporate Development & Sustainability Manager, Siemens

Siemens is a technology company focused on industry, infrastructure, mobility, and healthcare, creating technologies and transforming the industries that form the backbone of economies.

**Simon Juul Toft**, Lead Specialist—Green Procurement, Ørsted

Ørsted, the world's most sustainable energy company, is recognized as a global leader on climate action and takes tangible action to create a world that runs entirely on green energy.

**Thomas Udesen**, Chief Procurement Officer, Bayer, and Co-founder of The Sustainable Procurement Pledge (SPP)

Bayer is a global enterprise with core competencies in the Life Science fields of health care and agriculture. SPP brings together Procurement professionals to ensure responsible supply chains.

We would like to thank Michaela Bürger, supported by Nicole Ludwig, from Michaela Bürger Consulting for co-authoring on the substantial part of leadership and how to create a movement and drive transformation in the organization.

We also want to thank Max Bachmeier for visualizing our thoughts to ensure that our message gets across.

Special thanks go to Felix Simmons for his continuous support and for keeping us on track throughout the writing process.

Last, but not least, our thanks go to our editor Gary Walder whose support, creativity, input, and passion for sustainability played a significant role in shaping the narrative of this book.

# About the Book

## How to Read This Book

Our main goal was to combine theory with practical input to make this book become a "hands-on guide" for procurement specialists. Thereby, we enriched theoretical backgrounds with practical input from sustainability-related consulting projects and examples/opinions from expert interviews.

The book is structured as follows:

Chapter 1 aims to highlight the importance of sustainability and to assess the current situation while taking environmental, social, and corporate responsibility into account.

Chapter 2 continues with addressing the question of why it is important to act now and how you as procurement enthusiasts can make a difference. This chapter really sets the procurement in the spotlight and describes a holistic sustainability framework to get started.

Chapter 3 continues by providing a step-by-step approach to define your individual purpose, to assess your organization's current situation, and to set sustainability-specific targets and ambition levels.

Chapters 4–6 then address three key areas of action: (1) How to anchor sustainability into procurement's operating model, (2) How to set category work and suppliers in motion, and (3) How to create a movement and to drive transformation?

As a conclusion, Chap. 7 summarizes the book's key takeaways and will draw our picture of how sustainability in procurement will look in 2030.

Each chapter closes with a summary and key takeaways.

As we perceive these chapters to be a consecutive sequence of steps, a graphical illustration or "scribble" (see below) was prepared to provide general guidance on where to allocate the respective chapter into the overall sustainability journey. You will find a part of the illustration below at the beginning of each major chapter—hopefully helping you, the reader, grasp the key concepts before diving into the details.

To ensure easy reading, we decided to add textboxes as a design element to highlight quotes, best practices, and knowledge insights. Each chapter starts with a concise introductory quote taken from one of the expert interviews. Further quotes are added throughout the chapters to provide experts' evidence on our thoughts and opinion. These quotes are marked with the words "Sustainability in action."

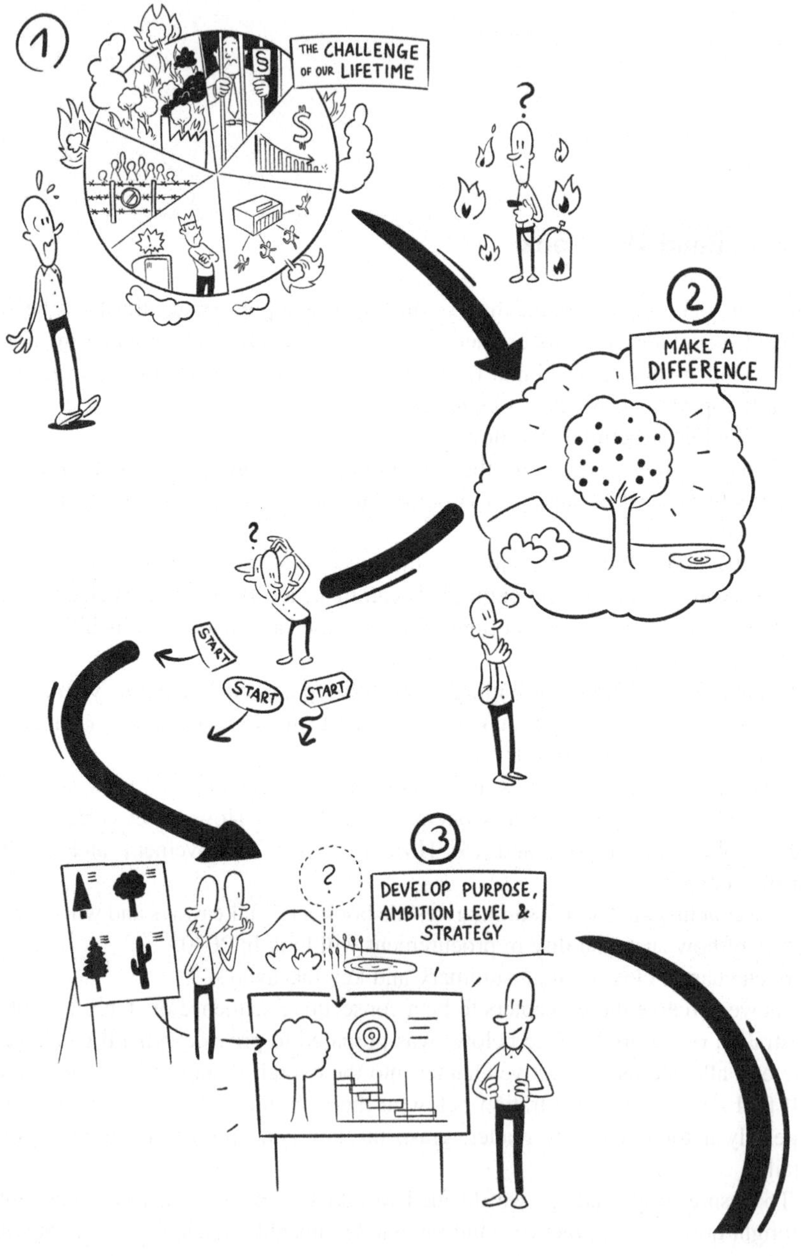

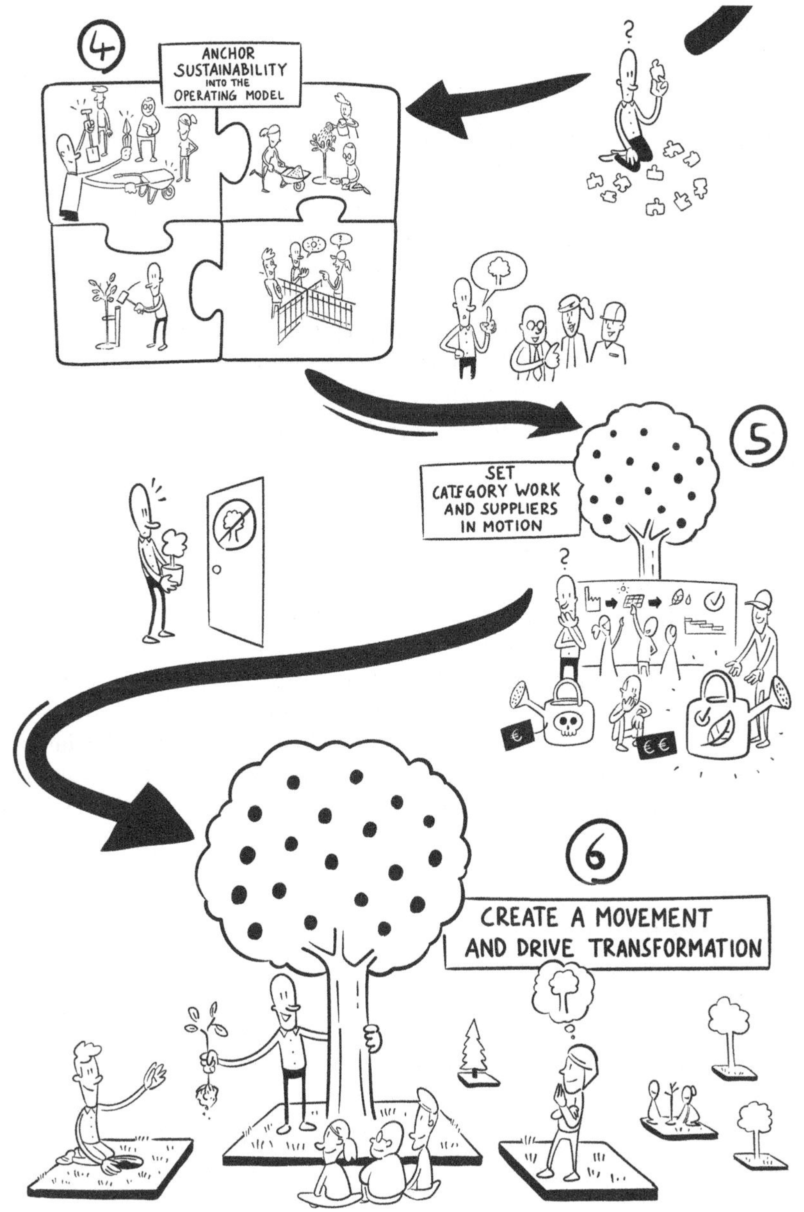
4
ANCHOR SUSTAINABILITY INTO THE OPERATING MODEL
5
SET CATEGORY WORK AND SUPPLIERS IN MOTION
6
CREATE A MOVEMENT AND DRIVE TRANSFORMATION

## Overview of Terminology

Finally, it seems that sustainability is being taken seriously. At the time these words were written, a Google search for "sustainability" resulted in a staggering 4.04 billion results on 08 December 2021, compared to 1.15 billion results on 12 July 2021. It is clearly a widely used word, and its apparent status as a contemporary buzzword might suggest that it is a new idea, but this is not the case. Unfortunately, it has taken more than 30 years for the word, or more precisely the concept, to become a frequent topic of conversation among friends, in news outlets, and in the boardroom where it is now looked upon as the new frontier for business (Billman, 2020).

The purpose of this chapter is to make it clear what we mean when we talk about sustainability or sustainable development. First, we will look at what is perhaps the most widely used definition of sustainable development—and the one we will use throughout this book—before looking at some of the more common sustainability frameworks.

## There Is Only One Future

The way we talk about sustainability today and what it entails has its origins in *Our Common Future,* a document written by the United Nations World Commission on Environment and Development, in 1987. It was in *The Brundtland Report*, as it became more commonly known after Gro Harlem Brundtland, who was the Commission's chairwoman at that time, that the first widely accepted definition of sustainability appeared: "Sustainable development is development that meets the needs of the present without compromising the ability of future generations to meet their own needs."

## The Triple Bottom Line

There are many different frameworks used when talking about sustainability but what they all have in common is that sustainable development involves consideration of not just the natural environment, but also stresses the need for social equity and economic improvement, or close variations of these three.

For example, the *triple bottom line* (TBL) is an accounting framework for sustainable business practices where *people, planet*, and *profit* represent *the social, environmental (or ecological)*, and *financial* performance of a company (United Nations Department of Social and Economic Affairs, 2021).

In a business context, no explanation is needed for *profit*. The same is not true for environment, or *planet*, which encompass pollution, eco-friendliness, carbon emissions, energy efficiency, and the six "Rs": refuse, reduce, reuse, repair, recycle, and rot. The social or *people* element refers to topics such as worker safety, fair labor practices, diversity, inclusivity, social engagement, human rights, fair compensation, and risk sharing.

## Corporate Governance Versus Corporate Social Responsibility

Corporate governance (CG) and corporate social responsibility (CSR) are also concepts that are linked with sustainability. Both CG and CSR are ways of ensuring that business activities are carried out in accordance with company objectives but it's important to distinguish between the two and what they represent.

Broadly speaking, CG is a blueprint for ensuring accountability, transparency, and compliance with due regard to a company's key stakeholders. It is an umbrella term that incorporates the legal, cultural, and institutional arrangements that determine what publicly traded corporations can and can't do and how controls are applied. These are mandatory controls that are aimed at internal stakeholders.

CSR is more a form of self-regulation that determines how companies integrate social and environmental concerns in their business operations and in their interactions with their stakeholders.

## The Greenhouse Gas Protocol

With its origins in the late 1990s, the Greenhouse Gas Protocol (GHG) (GHG Protocol, 2021) is another commonly used sustainability framework but, as the name suggests, its focus is specifically on greenhouse gas emissions. It provides comprehensive global standardized frameworks to measure and manage emissions from private and public sector operations and value chains, as well as mitigation actions. According to its website, it "enables companies and cities to develop comprehensive and reliable inventories of their GHG emissions, and helps countries and cities track progress toward their climate goals." In 2016, more than 92% of Fortune 500 companies were using the GHG Protocol to account for their greenhouse gas emissions.

## The 17 Sustainable Development Goals

A more recent and highly comprehensive approach to sustainability comes from the 2030 Agenda for Sustainable Development (United Nations Department of Social and Economic Affairs, 2021), adopted by all United Nations member states in 2015. This provides "a shared blueprint for peace and prosperity for people and the planet, now and into the future" (United Nations Department of Social and Economic Affairs, 2021). At its heart are the 17 Sustainable Development Goals (SDGs), which are an urgent call for action by all developed and developing countries to work together and "recognize that ending poverty and other deprivations must go hand-in-hand with strategies that improve health and education, reduce inequality, and spur economic growth—all while tackling climate change and working to preserve our oceans and forests."

Although the 17 SDGs were originally targeted toward nations, they are increasingly being adopted and adapted by corporations. Forward-thinking business leaders are aware that, compliance requirements aside, sustainability is essential for long-term competitiveness.

**Fig. 1** The United Nations 17 Sustainable Development Goals (United Nations, 2021)

### Sustainability in Action

> Due to the global climate emergency, we are facing, we developed our sustainability strategy with a strong focus on the environment. However, environmental, social, and economic issues cannot be solved in separation. There's a reason why, in 2015, the United Nations defined 17 interlinked Sustainable Development Goals to achieve a better future for all people and the planet—including goals like zero hunger, no poverty, quality education, life on land, climate action, responsible consumption and many more. To achieve a viable future for all people on a healthy planet we can't leave any problem unattended. Therefore, we're planning to address problems beyond environmental issues as well—again starting with a strong focus first until we're able to diversify and extend our efforts.
>
> **Nina Göntgen-Voss, Director of Sustainability, Forto**

## Our Understanding of Sustainability

There are slightly different definitions and frameworks dealing with the topic. We will use the original 1987 Brundtland definition (Sustainable development is development that meets the needs of the present without compromising the ability of future generations to meet their own needs). Throughout this book, we'll be linking this definition to different frameworks and tools that aim to apply it in practice.

Although there is mentioning of cities, regions, and countries in some of the frameworks listed here, we are specifically interested in how sustainability applies to companies and particularly to Procurement and Supply Chain organizations.

With those clarifications out of the way, the rest of the book will focus on how, not why, you should build a sustainable supply chain and procure goods and services sustainably.

# Contents

# About the Authors

**H&Z Management Consulting** H&Z was founded in 1997 by Rainer Hoffmann and Dr Thomas Zachau with the idea, that management consulting needs not just great concepts (Head), but also motivated clients (Heart) and sustainable implementation (Hand). Today the H&Z group has over 500 professionals and is among the leading consultancies for procurement and supply chain management. H&Z guides global players and SMEs in the fields of strategy, operations, and procurement—from optimizing processes, software solutions, management, and operational activities, to developing strategies that deliver growth.

H&Z's clients know that sustainability has become a strategic imperative. Beyond regulatory compliance, there are opportunities to make sustainability a competitive advantage and as COP26 has reiterated, it is time for companies to take responsibility.

As well as helping companies like Deutsche Bahn, Infineon, or Volvo Cars on their sustainability journeys, as a certified carbon-neutral company, H&Z is far along its own sustainability journey. Recently switching to exclusively green energy in the offices and committing to a 50% reduction in greenhouse gas emissions by 2030, this is certainly not the end of their journey. And the introduction of the SUSTAINX sustainability solutions platform only confirms this—it combines technological expertise and sustainability thought leadership to provide new innovative solutions and facilitates cross-industry exchange, to perpetuate corporate sustainability transformation.

Following the philosophy of "Consulting with Head, Heart, and Hand," H&Z works closely with its clients to create maximum impact and facilitate sustainable transformations and providing them with the capability to make sustainability an integral part of their operating model, of which in turn, procurement is a crucial component.

H&Z is part of The Transformation Alliance (TTA), an alliance of European-based consulting companies who believe in successful transformation through unlocking the full potential of all client employees. Located in all major markets, H&Z has over 1000 consultants working with their clients on a global scale.

**Stefan Aichbauer** is a Managing Partner at H&Z. Being an employee of the first hour, he has accompanied and driven H&Z on its growth path. While his duties include a broad range of responsibilities, his passion is on transformation and sustainability—combined they are the major challenge most corporates will face in the decade to come.

Stefan holds a Master in International Business and Economic Studies and a Master of Political Science from the University of Innsbruck. Recent accomplishments include Strategic Leadership in Action—a joint course from the military academy in Sandhurst and CASS Business School (City University of London).

**Martina Buchhauser** is a global leader with a deep understanding of the automotive industry and its transformation to new technologies and a responsible and $CO_2$-reduced business. She has held various executive positions at General Motors, MAN, BMW and until recently was the Chief Procurement Officer (CPO) of Volvo Cars and member of the management board. During her time at Volvo, she also served as a director of the advisory board of Geely Volvo Automobile Technology (Ningbo) Co. Ltd. Prior to Volvo, Martina was a senior vice president of purchasing and supplier network, interior, and electric/electronics, at BMW Group from 2012 to 2017.

Today, she is a non-executive director of various advisory boards and a senior advisor of boutique consulting firm, H&Z, specialized in strategy, procurement, and sustainability. Martina holds a Master of Science in management (M.Sc.) from Stanford University and a Bachelor of Science in business management (B.Sc.) from Lawrence Technological University.

**Michaela Bürger** (Guest author) gained more than 30 years of professional experience in operational and strategic management functions. Until 2008, Michaela led and designed the worldwide human resources development of Siemens AG together with the CEO. Since 2009, she has been advising on all challenges of talent and executive development management with a team of top-class experts. Her strengths lie in the development of integrated concepts for strengthening employee and leadership competencies in a digital world. Her thesis is that the quality of leadership behavior has a significant impact on corporate value. Based on this key thesis, she develops organizations in a goal-oriented manner.

In addition to the top resource of the individual, it is important to her that the visionary orientation results in an action-guiding version for all managers, which is consistent with the common understanding of leadership, and that there are implementable strategic thrusts for the development of the company as a whole.

**Agnes Erben** is a Partner at H&Z and manages projects to professionalize procurement organizations. Agnes started her professional career as a strategic purchaser for production material at MTU Aero Engines. There she got to know strategic, tactical, and operational challenges as well as her passion for procurement. She graduated from business administration, inter alia with a focus on materials management. Additionally, she holds a Master of International Business and Law from the University of Sydney.

In a previous publication, Agnes co-authored a practitioner's guide about "Procurement 4.0," describing a framework to embed digitalization in Procurement and master such challenges. In her view, Procurement can play a significant role in contributing toward Sustainability. Her work aims to link the two worlds of Procurement and Sustainability.

**Nicole Ludwig** (Guest author) has been working in the field of human resources development for 25 years. She knows both business and consulting and has held key positions in both organizations. Her focus is on co-designing integrated management concepts, developing IT-supported solutions, and implementing change projects based on networked and practical thinking.

She has a very good feel for people—in the selection of top executives as well as in coaching and mentoring. She manages transformation processes with the necessary care and with a strong, resilient relationship with everyone involved in the project. She is convinced that behavior makes the difference.

**Sven Steinert** is a Project Manager at H&Z with focus on supporting companies to shape a sustainable tomorrow. Prior to consulting, Sven was part of the business development team for renewable energy fairs at Messe Düsseldorf GmbH. At that time, he established the Energy Storage events in Düsseldorf and Tokyo. Sven holds a MSc in Management from Cass Business School, London with a major in corporate strategy.

In 2019, Sven co-founded PHOENX, a start-up producing travel fashion made from recycled and recyclable materials, where he is responsible for all supply-chain related activities. Moreover, he is leading SUSTAINX, a web-based platform initiated by H&Z and IntegrityNext, that provides access to sustainability solutions and that facilitates exchange among sustainability experts.

Sven sees sustainability as the essence to gain long-term competitive advantage and defines Procurement as the key contributor.

**Detlef Tietze** is a Partner at H&Z and responsible for the Sustainability Practice. He started in banking, but quickly moved to the consulting space. In over 16 years with H&Z, he managed many strategy and procurement projects. Today, he supports companies to build Sustainability strategies, to develop measures to deliver to the targets, and to anchor Sustainability in all aspects of the operating model. From his point of view, it's #TimeToAct.

Detlef holds a Ph.D. in Economics and a master's degree in business administration from the University of Bayreuth. To stay tuned, he participated in management courses at European Business School Oestrich-Winkel, HPI Hasso-Plattner-Institute, INSEAD and he holds a certificate in Business Sustainability Management from the University of Cambridge Institute for Sustainability Leadership.

**Emilia Wiking** is a Project Manager at H&Z with a focus on procurement and sustainability topics. Although an engineer by education, procurement has always had a focus in her career that started off within category management at A.P. Møller-Mærsk in Shanghai and continued to other procurement roles in China, Europe, and the USA. As a Swedish national, the sustainability mindset has been present since early days and she got to put it to practice when she joined Total's renewable energy team in Paris. Since then, she has been dedicated to helping companies improve their procurement practices in a sustainable way through her work at H&Z where she is a member of both the procurement and sustainability practice.

# Abbreviations

| | |
|---|---|
| (S)CoC | (Supplier) code of conduct |
| 3TG | Acronym for tungsten, tin, tantalum, and gold; refers to conflict minerals |
| CCP | Central clearing counterparty; takes on counterparty credit risk during a transaction. |
| $CO_2e$ | Carbon dioxide equivalents; measure indicating the amount of $CO_2$ needed to have same effect on the climate as another pollutant. |
| CSR | Corporate social responsibility |
| EFRAG | European Financial Reporting Advisory Group; the European voice on financial reporting discussing current reporting issues and major proposals to European financial reporting systems. |
| ESG | Environmental Social Governance; three key areas to focus on when discussing sustainability of businesses. |
| GHG | Greenhouse gases; any such emissions from carbon dioxide, methane, nitrous oxide, and F-gases summed up and measured in metric tons of carbon dioxide equivalents ($CO_2e$), where "equivalent" means "having the same warming effect as $CO_2$ over a period of 100 years" GHG Protocol, 2021. |
| GRI | Global Reporting Initiative; international standards organization for Sustainability impacts (Climate, human rights, and corruption). |
| IPPC | The Intergovernmental Panel on Climate Change is the United Nations' body for assessing the science related to climate change. |
| PESTEL analysis | Analysis from political, economic, social, technological, environmental, and legal perspectives |
| PPP | People, planet, profit; used in triple-bottom-line accounting (see below) |
| SASB | Sustainability Accounting Standards Board. Its standards enable businesses to identify, manage, and communicate financially-material sustainability information to their investors. |

| | |
|---|---|
| Scopes 1–3 | Scope 1 emissions are direct emissions from company-owned or controlled sources.<br>Scope 2 emissions are indirect emissions from the generation of purchased energy.<br>Scope 3 emissions are all indirect emissions (not included in scope 2) that occur in the value chain of the reporting company, including both upstream and downstream emissions. |
| SDG | Sustainable Development Goals are 17 goals defined by the United Nations as part of the 2030 Agenda for Sustainable Development. |
| TBL | Triple Bottom Line, also known as 3BL stands for people, planet, profit and is an accounting practice that places equal emphasis on social, environmental, and economic elements of business activities. |
| TCFD | Task Force on Climate Related Financial Disclosures to improve sustainability-related financial reporting. |
| TCO | Total Cost of Ownership refers to the cost of purchasing and operating an asset/the lifetime cost of an asset. |
| TVO | Total Value of Ownership refers to the opportunity cost of ownership, i.e. the advantage vs cost. |
| WFE | The World Federation of Exchange; global industry group representing exchanges and clearing houses (CCPs). |

# Part I

# Setting the Scene

# The Challenge of Our Lifetime

# 1

**100 s to Global Catastrophe**

In June 1947, a group of nuclear scientists from the former Manhattan Project launched the *Bulletin of the Atomic Scientists*, a magazine that monitors new developments in the life sciences and technology that could inflict irrevocable harm to humanity. That first issue had a simple image on its cover, a clock with its hands set at 7 min to midnight, that was the Bulletin's visual metaphor to warn how



S. Aichbauer et al., *Responsible Procurement*, Management for Professionals,
https://doi.org/10.1007/978-3-030-98640-7_1

close humanity was to annihilation—midnight—due to the danger of nuclear war (BBVA, 2021). Since then, the so-called Doomsday Clock has served as a warning and reminder of the global existential threats created by humanity against itself, which today also includes climate change and global pandemics. Over the years, its hands have been moved forward or backward depending on the intensity of the risk. Since 2020, the clock has been set at only 100 s to midnight, the closest it has ever been to the end in its entire history. The climate crisis was listed as one of the main reasons why we are now closer to the end of the world than ever before (Mecklin, 2021).

However, procurement can play an important role in moving the hands of the Doomsday Clock further back and pulling humanity away from the brink of collapse.

The clock is ticking, so it is time to change the way we treat the planet and all its ecosystems. It is time to change the way some people are abused for the sake of lowering the cost of consumer goods. It is time to change the way we live, time to change the way we do business, and time to put sustainability on top of the agenda when making supply chain and procurement decisions.

Sustainability, once a niche concern and away from the limelight for many people, is now well and truly the focus of mainstream media news headlines. This should add significant weight to the calls for action coming from governments, NGOs, consumers, and investors, who will in turn make sure that these calls are heard loud and clear in boardrooms across the world. There have always been progressive Chief Executive Officers and companies that have seen the opportunities present in sustainability. Now it is time for all companies to pay attention to what is going on around them. There is a seismic shift in favor of "accelerating sustainable solutions to all the world's biggest challenges—ranging from poverty and gender to climate change, inequality and closing the finance gap." The decade of action, a United Nations-backed campaign to transform our world within 10 years, appears to be underway (UN SDGs, 2021). There are of course challenges in the way, but the case for business sustainability is strong, and it will not go away. It is a case of ignore it at your peril.

## Sustainability in Action

Some KPIs are important: cost per kwh, quality, capacity ... but so is sustainability. When we first started to talk about this, no one listened. Four years later it has finally become important for companies and also for German customers. In my experience, its best not to start with compromises—if you cannot build a factory under similar $CO_2$ conditions as in Sweden, do build a second factory in a country with a high $CO_2$ energy mix. To produce truly green batteries, we need to manufacture them using renewable resources, such as hydropower, combined with a low carbon footprint around.

When we started to look at Europe, we talked with industry professionals, and knowing that they demand just-in-time concepts, there would have to be regionalization. We knew there was a case for building up capacities right here. We made the European Commission aware of the fact that Europe would have its own ecosystem, and that this would have a huge impact on R&D, equipment, component supply development, universities, and that the impact on the economy would be so much bigger if they were to support this. The European Battery Alliance was created out of this in 2017 and now has 500 member companies.

Now the focus is on the supply chain and how to build the competencies in R&D, manufacturing, and so on in a short period of time. There are some important topics that we're working on together with the EC and its European members. We're looking to find cobalt from non-Congo sources, and this could mean new ways of working with the raw materials and minerals. Again, to reduce the overall carbon footprint, we should build strong supply chains within Europe. Developing European machine builders and equipment manufacturers is also important for strategic competitiveness.

And let's not forget about recycling and the circular economy. These are vital parts of green manufacturing.

**Peter Carlsson, Co-founder and Chief Executive Officer, Northvolt**

Northvolt is striving to produce the greenest battery in the world with a minimal carbon footprint and the highest ambitions for recycling to enable the European transition to renewable energy.

## 1.1 The Environmental Challenges: Less Hot Air, Fewer Emissions, More Action

It was an event marked by stark contrasts, and perhaps the period from October 31 to November 12, 2021, will be remembered as a defining period in the future of planet Earth. Inside the Cop26 venue, there was an atmosphere of quiet and diplomatic deliberation as environment leaders from countries around the world debated how to progress towards the 1.5 °C target, which was the ultimate aim of the gathering. They seemed to be content with the outcome.

Outside, the huge crowds who had gathered in the streets around the SEC Center in Glasgow were less restrained as they loudly protested what they saw as a lack of progress, action, and commitment emerging from the conference to tackle the climate crisis. Greta Thunberg joined hundreds of other protestors who made their feelings quite clear as they chanted, "You can shove your climate crisis up your arse," in response to what they saw as too much "blah, blah, blah" coming from the politicians inside the venue and not enough concrete action (Gomez, 2021).

### Controversial Wordplay: Out Versus Down

Back inside, in what was described as "a dramatic 11th-hour decision, portrayed by green activists as a devastating blow to the success of Cop26," Bhupender Yadav, the Indian environment minister, read out a new version of the Glasgow pact that used the watered-down commitment to "phase down" coal rather than "phase out" (Ellis-Petersen, 2021).

Ironically, as Mr. Yadav read these controversial words, back in India, the country's capital city was suffering from the worst air pollution in more than 6 years, forcing the government to close schools and to warn people to stay indoors (BBC, 2021). Although crop burning and the smoke from the annual Diwali firework celebrations were initially blamed, a week later, the government ordered the temporary closure of five coal-fired power stations around New Delhi as part of its drive to combat life-threatening air pollution (CNBC, 2021).

## A Big Step Forward on Fossil Fuels

Although the precise implications of phase out versus phase down will be a contentious issue, this still represented progress. Since the Kyoto protocol was signed in 1997, there has been no reference to cutting out fossil fuels in a Cop decision, usually due to pressure from the major fossil fuel producers and users, who in the past have had the power to influence the agreements. The inclusion of such a commitment in the Glasgow outcome was thus a major step (Harvey, 2021).

## The Climate Challenge Is Very Real, and It Is Here

COP26 came towards the end of a year that brought home the harsh realities of climate change. In August 2021, the Intergovernmental Panel for Climate Change (IPCC) published its sixth report into the state of the climate, and it paints a grim picture (Masson-Delmotte et al., 2021). It effectively announced "a code red" warning for our species, saying there is now unequivocal evidence showing that greenhouse gas emissions and other unsustainable human activities were propelling us towards a calamitous and fiery future triggered by extreme climate change (McKie, 2021).

Examples of those extreme weather were on full display in Europe in 2021. In August, wildfires swept out of control across Greece, Turkey, and Sicily. The latter was also scorched by extreme temperatures, with the mercury hitting a high of 48.8 °C, Europe's highest ever temperature (Le Page, 2021). In July, flash floods caused huge loss of life and massive destruction in parts of Germany, Belgium, the Netherlands, and Switzerland. In June, Moscow was hit by a heatwave, with the temperature reaching a 120-year record high. In the Czech Republic, five people died and more than a hundred were injured when a tornado and hailstones the size of tennis balls hit the country's south-east. Earlier in the year in April, Arctic air swept across the continent, bringing record low temperatures and leaving French winemakers counting the cost of "the worst frost in decades." At the beginning of March, Germany recorded its sharpest temperature change in a week, going from −23.8 to 18 °C (Andrews, 2021). All of this in just a 6-month period and only in Europe. The situation was not much better in other parts of the world.

## A Sustainable and Green Recovery Is Called for

While all this was happening, we were still in the grip of the COVID-19 pandemic, which for many people was already a wake-up call for a renewed focus on sustainable development goals. In an interview with Deutsche Bank, scientist Richard Florizone, President and Chief Executive Officer of the International Institute for Sustainable Development, said he was seeing "growing calls across the world for a green recovery" from the pandemic and that the economic stimulus will have to be sustainable (Florizone, 2021).

That green recovery is something that all companies can be part of. The decision to act now, to re-write business strategy that has climate considerations and emissions reductions as central features, will pay dividends going forward. Thankfully, many progressive companies are already doing exactly this. For those that are prone to procrastination, the increasing number of environmental rules and regulations that are coming into play, as discussed in Sect. 1.3, will leave them with little choice but to toe the line.

## 1.2 The Social Challenges: Advancing Sustainable Development Through Human Rights

Concerns about climate change have dominated the headlines, but there is more to sustainability than extreme weather events—the social or human rights element is just as important. In fact, the environment and human rights are not distinct from each other, but part and parcel of the same issue, according to the UN Environment Program: "Human rights and the environment are intertwined; human rights cannot be enjoyed without a safe, clean and healthy environment; and sustainable environmental governance cannot exist without the establishment of and respect for human rights. This relationship is increasingly recognized, as the **right to a healthy environment enshrined in over 100 constitutions**" (UNEP, 2021).

Human rights create conditions essential for sustainable development, which is why the UN's 2030 Agenda recognizes "that inclusive and participative economies, and societies in which government is accountable, achieve better outcomes for all people, leaving no one behind" (UNHR, 2021). And yet, behind the scenes, human rights abuses and ill treatment of workers are still major concerns and more needs to be done to tackle this issue.

### Holding Companies to Account

In the past, too many companies from developed nations have turned a blind eye to what goes on in the factories and sweatshops in developing nations that are sub-contracted to supply their customers with cut-price smartphones, sneakers, or luxury handbags.

The infamous collapse in 2013 of the Rana Plaza building in Bangladesh that housed five garment factories making clothes for well-known fashion brands, when over 1100 workers died and a further 2000 were injured, is a tragic example of how competition for low prices and quick turnarounds can take priority over human rights. This was made all the worse for the fact that while some companies that made their clothes in Rana Plaza stepped in to pay compensation straight away, others only reluctantly paid after a 2-year legal battle. Other clothing companies tried to hide the fact that they had garments made in the building in a cynical bid to avoid compensation payouts and reputational damage to their brands (Hashad, 2015).

## Sustainability in Action

As you know, the responsible sourcing world has evolved from narrowly looking at conflict minerals, which often used to be portrayed as an electronics industry issue, to looking at the risk exposure of a broader set of raw materials underpinning the emerging electric vehicle economy. Human Rights risk management requirements are now complemented by additional risk criteria, for example, environmental transgressions or the specific issue of the exploitation of indigenous peoples' rights. Our aim has always been to create transparency for our customers by enabling them to understand how exactly their supplier ecosystem links them to specific risks in jurisdictions around the world. We provide the insights into the level of risk exposure associated with specific companies in different value chains and different geographies, and we show our customers that these aren't just some far-away issues that they can't do anything about. To the contrary, they are manageable risks with concrete solutions. Engaging the suppliers and supporting the implementation of specific corrective actions is critical to make progress. Measuring this progress and reporting on it is how you demonstrate continuous improvement. It's that demonstration of positive change what instills trust in consumers and stakeholders. That was really the starting point for us to effect change.

What we've seen particularly over the last five years was a surge of interest and acknowledgement by major brands in the auto sector, in the ICT (Information and Communication Technology) space or by retailers in general that there are problems in their raw material supply chains. Prior to that many were resistant to act because either they lacked the insights into how they are linked to a specific high-risk company or issue, or they feared acting would potentially open a can of worms and overwhelm them ... thinking that by beginning to act they were in some way "admitting" culpability. So, there were really strong forces against action and now this trend is changing.

**Dr Nicholas Garrett, Co-founder and Chief Executive Officer, RCS Global Group**

RCS Global Group is a proven leader in data driven ESG performance, responsible sourcing, and responsible mining assurance.

Apple and its manufacturing partner in China, Foxconn, were accused of a litany of worker violations, including withholding bonus payments, rolling back safety training, and employing more temporary workers than China's laws allow. Although Apple denied most of the allegations, Foxconn acknowledged that a review of its operations "did identify some workforce compliance issues" (Albergotti, 2019).

In the 1990s, Nike shocked its customers around the world when it was revealed that Indonesian and Vietnamese workers in the factories used to manufacture the company's footwear were routinely subjected to abusive labor practices. The allegations of abuse included stories that a Vietnamese sub-contractor ran women outside until they collapsed for failing to wear regulation shoes. Customers staged embarrassing public protests at the Olympics and at Nike stores and many people began boycotting the brand. It took more than 7 years—and hundreds of millions of dollars—to repair the damage to the brand. Ironically, Nike only found favor again with its customers when it became more transparent about the situation with its suppliers and started to respect the workers' rights in its sub-contracted factories (Lutz, 2015).

## Doing the Right Thing Is Not as Easy as It Seems

Another rapidly evolving area of concern over human rights abuses stems from the boom in demand for electric vehicles, particularly in Europe. This is a good thing because it means less air pollution and cleaner (European) cities. However, this has triggered a mad scramble for more cobalt, an essential ingredient of modern batteries. The problem is that more than 70% of the world's cobalt is produced in the Democratic Republic of the Congo (DRC), and 15–30% of the Congolese cobalt is produced by artisanal and small-scale mining. Human rights abuses are rampant in these mining operations and are particularly high in artisanal mines in the DRC, a country weakened by violent ethnic conflict, Ebola, and high levels of corruption. Child labor, fatal accidents, and violent clashes between artisanal miners and security personnel of large mining firms are recurrent. Simply shutting down these mines, or removing them from the cobalt supply chain, is not feasible if for no other reason than they are a lifeline for millions of Congolese who live in extreme poverty. Instead, companies committed to setting up responsible cobalt sourcing practices to take responsibility for addressing these human rights violations (Campbell, 2020).

### Sustainability in Action

It's a process that goes beyond just ensuring that there are no kids on the mine. It's a process that takes into consideration that the kids are on the mines for poverty-driven reasons, because their families need all hands-on deck to get food on the table. So, it means that you don't solve the issue necessarily by simply ensuring that there's not a single kid on the mine. That's important, because you don't want them to be working in a hazardous condition, but at the same time you can't just put them outside of the fence without something to eat. This is where UNICEF comes in with their expertise of working in local communities and driving real action within the communities, that basically offers these kids an alternative vision of the future. That shows how our presence in the Democratic Republic of Congo is leveraged by other actors as well. We can be on the mine and show there is a child, we can identify an issue of child labor and together with partners we can start to move them through a process that actually has a more long-term and more sustainable impact.

**Dr Nicholas Garrett, Co-founder and Chief Executive Officer, RCS Global Group**

RCS Global Group is a proven leader in data driven ESG performance, responsible sourcing, and responsible mining assurance.

## Workers' Rights: Getting Better but Far from Good

According to Human Rights Watch (HRW), there have been some improvements to workers' rights, especially in the wake of the Rana Plaza tragedy, but "millions of adults and children around the world suffer abuses as workers obtaining raw materials, toiling on farms, and making products for the global market. They are at the bottom of global supply chains, for everything from everyday goods like vegetables and seafood to luxury items like jewelry and designer clothing that end up on store shelves worldwide" (Roth, 2020).

This is why, according to HRW, there will be a growing trend of national legislatures requiring companies to live up to their responsibilities to workers, communities, and the environment. In July 2021, the German Supply Chain Act was passed into law, first time that the responsibility of German companies to respect human rights in global supply chains has been given a legal foundation. It follows the Modern Slavery Act in UK, the Child Labor Due Diligence Act in the Netherlands, and the EU Conflict Minerals Regulation. The table below lists human rights regulations from selected countries (Fig. 1.1).

## 1.3 Governmental Versus Corporate Responsibility: Do the Right Thing

Few businesspeople would disagree that keeping up with regulations is not the most enjoyable part of running a company., "Regulations" is something of a dirty word to many companies that operate in a free market economy because of the restrictions inherent in its definition. But they are a fact of life and compliance with the plethora of regulations that dictate what companies can and cannot do is important to avoid penalties due to malpractice.

In this respect, sustainability, including its human rights and environmental aspects, is well and truly in the spotlight and will be the subject of an increasing number of regulations enacted by different governments.

### Legal Versus Moral Imperatives

Leaving legal requirements aside for a moment, there is mounting pressure on companies to consider what is morally right and wrong in the context of sustainability and supply chain decisions. For example, how far should responsibility for sustainable business practices extend along the supply chain? According to the former German Industry and Federal Economics Minister Peter Altmaier, companies should pay attention to their product supply chains, but these due diligence obligations should not extend beyond the first tier (Klawitter, 2020). For leather for car seats, for example, BMW and Jaguar Land Rover would only have to trace their supply chain to the tannery in Veneto, Italy, a company that complied with the strict social and environmental standards of the two manufacturers. However, BMW, Jaguar Land Rover, and other European car manufacturers came under fire in 2020 when London-based NGO Earthsight, which specializes in human rights and environmental crimes, linked the illegal clearance of a South American forest inhabited by one of the world's last uncontacted tribes with some of these car manufacturers (Earthsight, 2020).

**Fig. 1.1** Human rights in the supply chain

Legally, the car manufacturers had done nothing wrong. But where do they stand morally? Can they simply wash their hands of the responsibility for the actual origin of the leather? Are they morally responsible for wrongdoing at the far reaches of their supply chain? There is a strong and compelling argument to say that even when a company is the last in the chain to receive components or finished materials, the responsibility for any lack of compliance should not be pushed onto lower-tier suppliers. BMW and Jaguar Land Rover are far from being the only two companies that face these questions of right and wrong. Irrespective of what the law says, consumers and investors still want them to do the right thing.

## Increased Legislation Is on the Horizon

It is precisely because businesses do not always do the right thing when it comes to sustainability that there is an increasing amount of legislation covering it. In March 2021, Germany's Federal Government launched the new Supply Chain Act, legislation to protect the rights of people who produce goods for the German market. The act was created to ensure that German companies comply with human rights for workers along their entire supply chain and that they set up grievance mechanisms and report on their activities. It comes into effect in 2023 for companies with a workforce of at least 3000 and later for companies with a workforce of at least 1000 (The Federal Government of Germany, 2021). France and The Netherlands have enacted similar laws, and other European countries will undoubtedly do the same.

## Climate Change Grabs the Headlines, but Human Rights Are Not Forgotten

In fact, on March 10, 2021, the European Parliament considered and adopted an outline proposal for the "EU Directive on Mandatory Human Rights, Environmental and Good Governance Due Diligence" (the Directive) following concerns that only one business in three is currently conducting appropriate due diligence measures with regards to its value chain. Although the Directive is not expected to come into force until late 2022 or early 2023, companies falling within its scope will need to begin preparations before the new requirements kick in. Any business operating in Europe, irrespective of place of registration, with more than 250 employees or a turnover of 50 million euros, will be affected (Da Costa & Marsh, 2021).

Further legislation followed in June 2021, when all European Union countries, with the exception of Bulgaria, approved a law to make the bloc's greenhouse gas emissions targets legally binding in a landmark bid to tackle climate change. The deal sets targets to reduce net EU emissions by 55% by 2030, from 1990 levels, and eliminate them by 2050. Other policies will stem from this, designed to reshape industry, energy, transport, and housing to emit less $CO_2$ (Abnett, 2021).

### More Stringent Reporting

There is also the EU Corporate Sustainability Reporting Directive (CSRD) that comes into play 2023, with mandatory reporting from 2026 onwards. It requires businesses with more than 250 employees to disclose information on the way they operate and manage social and environmental challenges and will cover more sustainability-related information than ever before, including sustainability targets and indicators, resilience against sustainability-related risks and climate scenarios, social considerations, as well as governance processes. The law has been put in place to help investors, civil society organizations, consumers, policy makers and other stakeholders to evaluate the non-financial performance of large companies and encourages these companies to develop a responsible approach to business (European Commission, 2021a).

### Carbon Emissions Outside Europe Are Part of the Clean Up

Climate change is a global problem that needs global solutions and as the EU raises its climate ambitions, it is making sure that EU companies are accountable for carbon emissions produced outside the bloc where less stringent environmental and climate policies prevail. EU companies found guilty of so-called "carbon leakage," where carbon-intensive production is moved abroad to take advantage of lax standards, or EU products are replaced by more carbon-intensive imports, will still have a price to pay. The Carbon Border Adjustment Mechanism (CBAM), introduced in July 2021, will equalize the price of carbon between domestic products and imports and ensure that the EU's climate objectives are not undermined by production relocating to countries with less ambitious policies (European Commission, July, 2021b)

The message could not be clearer—companies operating in the EU and sourcing from outside the bloc will soon have no choice but to clean up their businesses at home and their supply chains overseas.

### Corporate Sustainability Responsibility: Going Above and Beyond Compliance

These new laws and regulations enacted by governments could be viewed as a necessary evil to ensure that mandatory sustainability targets are met.

Or they could be seen as an opportunity.

Some forward-thinking, progressive companies are looking at the increasing number of sustainability-related legal obligations worldwide not as hindrances, but as a powerful force in driving sustainable business practices. *In the run up to COP26, 778 businesses representing US$2.7 trillion in annual revenue and employing 10 million people urged G20 leaders to take more action to limit average global temperature rise to 1.5°C* (We Mean Business Coalition, 2021).

In an open letter to the G20, this coalition of progressive companies stated: "There are great opportunities to be harnessed through a carefully managed transition. Our businesses recognize the benefits of climate action. The right policy decisions taken today can drive further investments and spur business decisions in favor of climate solutions across G20 countries. We can build stronger, just, and more resilient economies: bringing prosperity and creating decent jobs while protecting health and the planet" (ibid).

The message coming from this is that regulations are a good starting point, but there needs to be unprecedented cooperation between governments and corporations to jointly embrace responsibility and accelerate the race to zero.

In an interview for this book, Dr. María Mendiluce, Chief Executive Officer at the We Mean Business Coalition, said she was fully behind the need for strong legislation, open reporting, and collaboration between countries and the businesses operating within their borders.

### Sustainability in Action

One thing is very clear to me: In most cases, when it comes to companies publicizing their commitments and tracking their progress, they tend to outperform their own targets. We need much more of that. Unfortunately, the opposite is true of some countries. But if we could get countries working closely with their domestic companies to achieve their targets, we could move from a system that's voluntary to one where it's mandatory for businesses to be net-zero and reduce emissions. We're not ready yet but in a few years, because things are getting tough with climate change, we'll get there. This means that it's better for companies to get ahead of the game now before they're forced to later.

**Dr Maria Mendiluce, Chief Executive Officer, We Mean Business Coalition**

We Mean Business Coalition is a global non-profit coalition working with the world's most influential businesses to take action on climate change.

## Collaboration, Not Competition, Is the Way Forward

The rules around collaboration, usually prohibited by EU and national laws because it restricts competition, are being relaxed where sustainability is concerned. In January 2021, the UK's Competition and Markets Authority (CMA) released guidelines to help businesses understand how to navigate sustainability agreements between businesses (including industry-wide initiatives) and achieve green goals without breaching competition law. This follows a similar initiative by the Dutch Competition Authority, which has provided guidance to help businesses self-assess the compatibility of joint sustainability initiatives with competition law, without the risk of fines for companies applying the guidelines in good faith (Gardner, 2021).

By removing this legal barrier, businesses are able to join forces to pursue environmental objectives together, for example through joint R&D of green technologies, a joint commitment to minimum environmental standards or through coordinating resources or joint purchasing. Together, these measures could help to create a level playing field, achieve economies of scale, and reduce investment costs.

Some major players are taking full advantage of this era of collaboration to find sustainable solutions to industry-specific, climate-related problems. Henrik Larsen from A. P. Møller-Mærsk spoke about the importance of collaboration to tackle sustainability problems.

### Sustainability in Action

The same reasoning applies to our decision to establish a research center, the Mærsk Mc-Kinney Møller Center for Zero Carbon Shipping. This not-for-profit, independent research and development center is using the synergies of cross-sectoral collaboration from the worlds of science, engineering, and business to accelerate the development and implementation of new energy systems and technologies that will decarbonize the global shipping sector.

**Henrik Larsen, Chief Procurement Officer at A. P. Møller-Mærsk**

A. P. Møller-Mærsk is an integrated container logistics company and member of the A.P. Møller Group. With a dedicated team of over 80,000 people, they enable global trade for a growing world.

Collaboration is the way forward.

### Sustainability in Action

At this point we see a collaborative approach as the best way to motivate our suppliers to join the same decarbonization journey as us. We're currently focusing on our strategic suppliers for the decarbonization initiative. Through these strategic suppliers we also work with our second-tier suppliers for key materials such as steel. Decarbonizing the supply chain of renewable energy is complex and requires a significant amount of innovation and collaboration within the industry and across sectors as it's also tied to the decarbonization of other key sectors, including heavy manufacturing and heavy transportation. As such, the core of Ørsted's decarbonization program since its launch in 2020 is to promote collaboration towards our common goal across our base of strategic suppliers in the most $CO_2$ intensive categories. To succeed in this goal, all key actors in the industry must play their part.

**Simon Juul Toft, Lead Specialist—Green Procurement, Ørsted**

Ørsted, the world's most sustainable energy company, is recognized as a global leader on climate action and takes tangible action to create a world that runs entirely on green energy.

## Sustainable Business Practices Are Finally Gathering Momentum

Legislation is necessary, but businesses should not wait for this to get started on their sustainability transformations. They should take the initiative and go beyond the baseline of compliance. By improving existing standards and practices, and by being accountable for their actions, businesses can get closer to achieving a more sustainable future, which is what customers, investors, and the world in general needs. Beyond this, the early movers will begin to reap the benefits of a sustainable approach to business: improved operations, enriched risk management, strengthened stakeholder relationships, costs savings, value creation, and improved market positioning.

### Sustainability in Action

I think procurement has an enormous responsibility to explore innovative ideas that will bring benefits to the company. Right now, the benefit is not very clear because there's no carbon pricing included. Ultimately, consumers are asking for reduced $CO_2$ footprints on products, so you'll have to trickle this down to the supply chain and make it happen. That is the future.

I also think that companies, boards, and CEOs and management teams will have $CO_2$ targets as part of their incentive and compensation schemes. This will be cascaded down through the organization so that procurement will have both an environmental cost and a financial incentive and they will optimize both. Once carbon pricing becomes part of the objectives of companies, they will have to include it as part of their procurement strategy.

**Dr Maria Mendiluce, Chief Executive Officer, We Mean Business Coalition**

We Mean Business Coalition is a global non-profit coalition working with the world's most influential businesses to take action on climate change.

## Consumers Want to Save the World

The environmental and social arguments of sustainability are compelling reasons for change in themselves, but they are not the only ones pushing companies to clean up their acts. There are compelling dollar-related reasons as well.

Consumers are driving the sustainability agenda through their buying behavior and are increasingly looking for brands "that operate according to their values, and prioritize employee wellbeing, sustainability, and helping the most vulnerable," according to a 2020 report published by Brandwatch (Brandwatch, 2020). The report looked at how consumers view the role and responsibilities of businesses and concluded that "people no longer feel it's acceptable for brands to be purely profit driven," and instead want to see the companies they buy from "taking action to improve the world we live in."

## Investors Demand Sustainable Returns

These high-level sustainability actions suggested by leading international intergovernmental organizations and NGOs, renewed governmental legislation, as well as changing consumer behavior have got the attention of investors who are increasingly conscious of where and how their money is invested. In a well-publicized letter, BlackRock, one of the world's biggest investment houses with more than US$8.7 trillion assets under management, told its clients that "because sustainable investment options have the potential to offer clients better outcomes, we are making sustainability integral to the way BlackRock manages risk, constructs portfolios, designs products, and engages with companies. We believe that sustainability should be our new standard for investing" (Fink, 2020).

## It Is Time to Act Now!

The scene has been set. Climate change is a real and ever-present danger to the world we know. It is an imperfect world, where profits and quick turnarounds have been favored above basic human rights, and where exploitation of natural resources takes place on a massive scale with too little regard for the long-term consequences. It could be argued that these consequences are beginning to bite now, as we see with the increasing frequency and severity of human-induced natural disasters. The Doomsday Clock is ticking, but we have a clear role in determining whether its hands move us closer to or further away from catastrophe.

## References


Abnett, K. (2021). EU countries approve landmark climate change law. *Reuters Sustainable Business*. Retrieved August 16, 2021, from https://www.reuters.com/business/sustainable-business/eu-countries-approve-landmark-climate-change-law-2021-06-28/

Albergotti, R. (2019). Apple accused of worker violations in Chinese factories. *The Washington Post*. Retrieved December 7, 2021, from https://www.washingtonpost.com/technology/2019/09/09/apple-accused-worker-violations-chinese-factories-by-labor-rights-group/

Andrews, M. (2021). Europe's unusually warm week breaks temperature records. *The Guardian*. Retrieved August 16, 2021, from https://www.theguardian.com/weather/2021/mar/04/europe-unusually-warm-week-breaks-temperature-records

BBC. (2021). Air quality: Delhi records worst November air in years. *BBC News*. Retrieved December 7, 2021, from https://www.bbc.com/news/world-asia-india-59486806

BBVA. (2021). The doomsday clock—75 years predicting the end of humanity. *Open Mind BBVA newsletter*. Retrieved December 10, 2021, from https://www.bbvaopenmind.com/en/science/environment/the-doomsday-clock-75-years-predicting-the-end-of-humanity/

Brandwatch. (2020). Six ways consumer mindsets have changed since Coivd-19. *Marketing Week*. Retrieved July 27, 2021, from https://www.marketingweek.com/six-ways-consumer-mindsets-changed-COVID-19/

Campbell, J. (2020). *Why Cobalt mining in DRC needs urgent attention*. Council on Foreign Relations. Retrieved December 12, 2020, from https://www.cfr.org/blog/why-cobalt-mining-drc-needs-urgent-attention

CNBC. (2021). India temporarily shuts five coal-fired power plants around New Delhi. *CNBC*. Retrieved December 7, 2021, from https://www.cnbc.com/2021/11/17/india-temporarily-shuts-five-coal-fired-power-plants-around-new-delhi.html

Da Costa, G. R., & Marsh, J. P. M. (2021). European Union moves towards mandatory supply chain due diligence: Start gearing up for new directive. *The National Law Review*. Retrieved August 16, 2021, from https://www.natlawreview.com/article/european-union-moves-towards-mandatory-supply-chain-due-diligence-start-gearing-new

Earthsight. (2020). *Grand Theft Chaco—The luxury cars made from the stolen lands of an uncontacted tribe*. Earthsight. Retrieved August, 16, 2021, from https://www.earthsight.org.uk/grandtheftchaco-en

Ellis-Petersen, H. (2021). India criticised over coal at Cop26—But real villain was climate injustice. *The Guardian*. Retrieved November 26, 2021, from https://www.theguardian.com/environment/2021/nov/14/india-criticised-over-coal-at-cop26-but-real-villain-was-climate-injustice

European Commission. (2021a). *Corporate sustainability reporting*. European Commission. https://www.ec.europa.eu/info/business-economy-euro/company-reporting-and-auditing/company-reporting/corporate-sustainability-reporting_en

European Commission. (2021b, July). *Carbon border adjustment mechanism: Questions and answers*. European Commission. https://www.ec.europa.eu/commission/presscorner/detail/en/qanda_21_3661

Fink, L. (2020). *Sustainability as Blackrock's new standard for investing*. Retrieved July 27, 2021, from https://www.blackrock.com/corporate/investor-relations/2020-blackrock-client-letter

Florizone, R. (2021). *The pandemic helps focus on global sustainability goals*. Deutsche Bank. Retrieved August 16, 2021, from https://www.db.com/what-next/responsible-growth/dossier-COVID-sustainable-development/pandemic-helps-focus-on-global-sustainability-goals?language_id=1

Gardner, J. (2021, March). *Sustainable collaboration*. Fieldfisher. Retrieved November 25, 2021, from https://www.fieldfisher.com/en/services/competition-and-consumer/competition-connect-blog/sustainable-collaboration

Gomez, D. (2021). Greta Thunberg mocks COP26 politicians at rally: 'Shove the climate crisis up your Arse'. *The Wrap*. https://www.thewrap.com/greta-thunberg-cop26-shove-the-climate-crisis-up-your-arse/

Harvey, F. (2021). Ratchets, phase-downs and a fragile agreement: How Cop26 played out. *The Guardian*. Retrieved December 6, 2021, from https://www.theguardian.com/environment/2021/nov/15/ratchets-phase-downs-and-a-fragile-agreement-how-cop26-played-out

Hashad, D. (2015). When workers die, no company can walk away. *Al Jazeera*. https://www.aljazeera.com/opinions/2015/4/24/when-workers-die-no-company-can-walk-away/

Le Page, M. (2021). Sicily hits 48.8°C, the highest temperature ever recorded in Europe. *The New Scientist*. Retrieved August 16, 2021, from https://www.newscientist.com/article/2286967-sicily-hits-48-8c-the-highest-temperature-ever-recorded-in-europe/

Lutz, A. (2015). How Nike shed its sweatshop image to dominate the shoe industry. *Business Insider*. Retrieved December 8, 2021, from https://www.businessinsider.com/how-nike-fixed-its-sweatshop-image-2015-6

Masson-Delmotte, V., et al. (2021). *Climate change 2021: The physical science basis*. Contribution of Working Group I to the Sixth Assessment Report of the Intergovernmental Panel on Climate Change. Retrieved August 16, 2021, from https://www.ipcc.ch/report/ar6/wg1/

McKie, R. (2021). It's now or never: Scientists warn time of reckoning has come for the planet. *The Guardian*. Retrieved 16 August, 2021, from https://www.theguardian.com/environment/2021/aug/15/its-now-or-never-scientists-warn-time-of-reckoning-has-come-for-the-planet

Klawitter, N. (2020). Deforestation for European car seats. *Der Spiegel*. Retrieved August 16, 2021, from https://www.spiegel.de/wirtschaft/leder-aus-paraguay-fuer-bmw-land-rover-und-co-abholzung-fuer-europaeische-autositze-a-dc0f3969-9ebb-4b66-982c-daee809e1d5b

Mecklin, J. (2021). This is your COVID wake-up call: It is 100 seconds to midnight. *Bulletin of the Atomic Scientists*. Retrieved December 10, 2021, from https://www.thebulletin.org/doomsday-clock/current-time/

Roth, K. (2020). *Holding companies to account: Momentum builds for corporate human rights abuses*. Human Rights Watch. Retrieved August 18, 2021, from https://www.hrw.org/world-report/2020/country-chapters/global-2

UNEP. (2021). *What are environmental rights?* United Nations Environment Programme. Accessed November 23, 2021, from https://www.unep.org/explore-topics/environmental-rights-and-governance/what-we-do/advancing-environmental-rights/what

UNHR. (2021). *Advancing sustainable development through human rights*. United Nations Human Rights. Retrieved November 23, 2021, from https://www.ohchr.org/EN/AboutUs/ManagementPlan/Pages/sustainable-development.aspx

UN SDGs. (2021). *Decade of action—United Nations sustainable development*. Retrieved July 23, 2021, from https://www.un.org/sustainabledevelopment/decade-of-action/

Unknown Author. (2021). *Greater protection for people and the environment in the global economy*. The Supply Chain Act (Lieferkettengesetz), The Federal Government of Germany.

Retrieved August 16, 2021, from https://www.bundesregierung.de/breg-en/federal-government/supply-chain-act-1872076

We Mean Business Coalition. (2021). COP26: Businesses urge world leaders to keep 1.5°C alive. *We Mean Business Coalition*. Retrieved November 23, 2021, from https://www.wemeanbusinesscoalition.org/g20-2021/

# Make a Difference

# 2

S. Aichbauer et al., *Responsible Procurement*, Management for Professionals,
https://doi.org/10.1007/978-3-030-98640-7_2

## 2.1 Common Excuses

Find fun in the subject. It's very meaningful. When someone is just starting out, I think the most important thing is to understand sustainability within the overall context of the business. This will show where the opportunities lie. At the end of the day, no company can ignore sustainability nowadays, so embrace it.

**Nils Eichberger, Vice President of Procurement, LBBW (Landesbank Baden-Württemberg)**

### "Killer" Phrases

The business case for sustainability is strong, as are the ethical arguments. If the combination of these was not enough, the increasing amount of sustainability legislation, pressure from investors, from employees, and from consumers are compelling. In other words, there are no rational reasons why sustainability should not be at the core of all companies. And yet there are still some common excuses that fuel procrastination and stand in the way of action.

Together with our colleagues from Nomura Research Institute Japan, we collected an elaborate list of frequently used excuses and possible replies. Nothing should stand in the way of a sustainability transformation.

**Sustainability is just another hype.**

This argument is heard less frequently but it still persists. With every extreme weather event close to our homes, this argument is rapidly becoming redundant. And consumers, employees, and investors are making sure that sustainability is not mere hype, but a concrete reason for change that will benefit everyone.

**Sustainability is only window-dressing and green washing. We should reduce or eliminate carbon dioxide, but there's no need to compensate for emissions.**

This is usually brought up to muddy the waters by adding an unnecessary layer of complication, and it is certainly true that eliminating $CO_2$ emissions in the first place is the ideal. Compensation at least puts a price tag on $CO_2$ and where there is a cost, there is a business case to mitigate that cost.

**Sustainability is all about additional costs. There are no gains.**

In some industries, this is currently the case, but when regulations and/or consumer demands kick in, those businesses that are ahead of the curve will prevail. Investors and owners increasingly see the benefits of investing in sustainable businesses, so sustainability is all about making businesses futureproof. Most managers look for suitable ways to monetize new business models while keeping additional costs to the minimum.

**Sustainability is too expensive.**

Yes, sitting and waiting is an option, but so is going out of business. There is clear evidence showing that customers want to buy from "green" companies and employees want to work for companies that are not just focused on the bottom line. A 2020 workplace survey found that 72% of multigenerational respondents were concerned about environmental ethics, with an astounding 83% of workers reporting their workplaces were not doing enough to address climate change (Acaroglu, 2020). When it is becoming increasingly difficult to hire and retain the right employees, top talents are prioritizing purpose over salary and opting to work for companies that have strong environmental credentials. Sitting on the fence where sustainability is concerned will have customers and employees looking elsewhere.

**Dealing with sustainability isn't necessary, we don't need to do this.**

Well, this book is certainly not for everyone, but surely saving the planet is. Everybody has the right to an opinion, but facts are facts. Perhaps it is time to step aside and be replaced by someone who is willing and able.

**We are running a business, not a charity.**

An attitude like that is not good for business. ESG compliance is no longer an option but a requirement and a sign of a healthy business. Reluctance to comply will lead to investors pulling their funds, customers shopping elsewhere, and employees walking out the door in search of a company that has a better attitude.

The economic factors of sustainable business are a key challenge. Becoming sustainable, or developing new sustainable business models, requires huge investments. Most managers are looking for suitable ways to monetize new business models while keeping additional costs to the minimum.

**It is a very complex issue, where do I start?**

There is no pre-set pathway leading to the right actions and there is a lack of solutions for some sustainability-related problems. The way forward is to look for solutions, develop new business opportunities and make the best use of what is already out there to find a path through the sustainability jungle.

**I don't have top management support.**

There are two options. One, sit out the storm and start grassroot initiatives and see where that leads. Two, leave and find a new workplace where care for people and planet are part of the core values.

**We don't have the resources.**

Start small and make the case for change and if there are no resources for scaling up, look again at the options when there is "no top management support." Search for allies within the company or in the wider ecosystem. Investors, customers, or suppliers can be great allies.

**We don't have enough data—the business case is not yet robust enough.**

Data is the lifeblood of many processes, but not every action requires data right away. Again, start small and scale while amassing the data that is still lacking. In addition, there are organizations or NGOs that can help to provide the databases. For example, South Pole is one such organization that helps companies to build $CO_2$ databases based on benchmarks from various industries. Classical should-costing tools can also provide the carbon footprints of products and can help to build the case for an improved carbon footprint.

### Sustainability in Action

One major challenge is how we can be sure that we really get what we need. We're reliant on external parties, tool providers, and our partners in the supply chain to provide us with information and data, but how can we be 100% sure that it's correct? The amount of time and resources that you need to invest also require careful consideration and planning.

**Henrik Larsen, Chief Procurement Officer, A. P. Møller-Mærsk**

A. P. Møller-Mærsk is an integrated container logistics company and member of the A.P. Møller Group. With a dedicated team of over 80,000 people, they enable global trade for a growing world.

## 2.2 Time to Shine for Procurement

I think procurement has an enormous responsibility to explore innovative ideas that will bring benefits to the company.

**Dr María Mendiluce, Chief Executive Officer, We Mean Business Coalition**

### A New Dawn for Procurement

It is time for procurement to come out of the shadows. It is time for procurement, too often considered the poor relation in comparison to the more dominant business functions such as R&D, manufacturing, or sales to show that it can spearhead real, transformational change that will benefit the entire company. It is time for procurement to be seen in a new light and we think the timing could not be better.

### Sustainability in Action

We're not there yet, but companies are going to reduce scope 1 and scope 2 emissions rapidly but they'll struggle to reduce scope 3 emissions. Then they'll look at procurement offices and ask 'what are you going to do about this?' Procurement leaders need to be at the forefront in the struggle to reduce scope 3 emissions—they need to know that this is coming so they should already get started.

**Dr Maria Mendiluce, Chief Executive Officer, We Mean Business Coalition**

We Mean Business Coalition is a global non-profit coalition working with the world's most influential businesses to take action on climate change.

In the past, procurement was viewed as the corporate underdog, overshadowed by its more glamorous corporate siblings such as sales, marketing, R&D, and manufacturing. It would not be the first choice for many young, talented businesspeople who are looking to develop their careers. But this point of view is rapidly becoming outmoded. To some extent, the COVID-19 pandemic accelerated interest in business sustainability. Numerous reports and studies have shown that the coronavirus has focused attention on the need to create a better, healthier world. For instance, a study by research group Kantar said that since COVID-19, sustainability was more of a concern for consumers than before the outbreak (Kantar, 2020). And

65% of respondents told a survey by pollsters Ipsos Mori "that it is important that climate change is prioritized in the economic recovery after coronavirus" (IPSOS Mori, 2020).

### Sustainability in Action

> I think that's where the procurement executives will take a bigger stance and this could be the inflection point for procurement sitting on the board. Procurement has the ability to help deliver sustainable business growth by uniting the environmental, social, and economic impacts of the organization and also by aligning its ecosystem with the strategy.
>
> **Mark Perera, Chief Executive Officer, Vizibl**
>
> Vizibl is a digital procurement platform that allows global organizations to unlock the true value of their supplier relationships.

There is unprecedented demand from consumers and investors for the business world to behave more responsibly, and an increasing amount of legislation to make sure companies do exactly that. This has put sustainability higher on the corporate agenda than it has ever been, and along with it comes procurement. With its huge responsibility for substantial parts of corporate spending and oversight of suppliers, procurement can and should be spearheading the push for sustainable business.

This is probably the most exciting and challenging time for procurement to prove how value can be created beyond pushing for price in the sourcing process. In response to the emphasis on sustainability, transformation processes are already underway in many industries to reshape procurement. Doing nothing is not an option—now is the time to put procurement on the map.

## Now Is the Time to Act

### Sustainability in Action

> Hardly any supply chain is "clean", and it is important not to be blinded by big statements that suggest that supply chains are. Instead, I think responsible sourcing and supply chain engagement has correctly been construed as a continuous improvement effort. I think it's the right position, that it acknowledges that this is an ongoing process that needs ongoing attention which, in practice, means board-level attention. Board-level attention can ensure companies move in a better direction, while these directional improvement efforts then also have a higher chance of being correctly resourced. But, at the same time, continuous improvement needs to be underpinned by data. You need to actually show a report that proves, based on verifiable data, that you are continuously improving.
>
> **Dr Nicholas Garrett, Co-founder and Chief Executive Officer, RCS Global Group**
>
> RCS Global Group is a proven leader in data driven ESG performance, responsible sourcing, and responsible mining assurance.

Admittedly, it would have been better for all of us if the business world had jumped onto the sustainability bandwagon a few decades ago with the same enthusiasm and vigor that we are seeing today, but that is a matter for history to decide. Or perhaps it is because there is now, for the first time, an alignment of factors that have made it abundantly clear to even the most hard-nosed cynic that this is the time to act. Intense consumer pressure, insistence from investors and the financial markets, increasing legislation, an impending climate catastrophe, and the expectations a new generation of professionals have of the companies they work for have created an unprecedented and powerful force for change.

## Consumers: Letting Their Fingers Do the Talking

Consumers around the world are making their opinions evidently clear. A May 2021 report released by the Economist Intelligence Unit (EIU), and commissioned by WWF, shows a staggering 71% rise in popularity of searches for sustainable goods over the past 5 years, with continuing growth during the COVID-19 pandemic (WWF, 2021). The report, *An Eco-wakening: Measuring global awareness, engagement and action for nature*, shows a surge in consumer clicks for sustainable goods in high-income countries such as the USA, UK, Germany, Australia, and Canada. However, it is not just in developed economies where consumers are looking for sustainable products. The trend is also accelerating in developing and emerging economies such as Ecuador (120%) and Indonesia (24%). The most dramatic growth in engagement and awareness has occurred in Asia, most notably in India (190%), Pakistan (88%), and Indonesia (53%). Clearly, people all over the world care about sustainability and that trend is growing, indicating that sustainable and planet-friendly products are no longer a niche market.

## Regulations: Do It Our Way or Not at All

Rising concerns about accelerating climate change have resulted in increasing regulations and a growing trend for litigation, especially around $CO_2$ emissions. Some of the litigation has directly cited value chains as areas of increasing risk and concern.

In a push to become the first climate-neutral continent by 2050 and as part of its European Green Deal, the EU has adopted ambitious new targets to curb climate change that will become legally binding. Under a new law agreed between member states and the EU Parliament, the bloc will cut carbon emissions by at least 55% by 2030, compared with 1990 levels (BBC, 2021). The new plan will require major overhauls of the energy and transport sectors as well as a huge push to renovate and retrofit buildings to make them energy-efficient and able to charge electric vehicles.

As part of this, a new EU law that came into effect in 2021 requires certain large companies to disclose information on the way they operate and manage social and environmental challenges, a ruling that at the time of writing applied to 11,700 large public-interest companies with more than 500 employees across the EU. A collaboration between the Global Reporting Initiative (GRI) and the European Financial Reporting Advisory Group (EFRAG) was established to develop reporting standards that will assist the European Commission in fulfilling its objective to increase corporate transparency in support of the European Green Deal (GRI, 2021).

## Sustainability-Based Litigation Is on the Rise

Around the world, climate change litigation continues to grow. Globally, the cumulative number of climate change-related cases has more than doubled since 2015. Just over 800 cases were filed between 1986 and 2014, while over 1000 cases have been brought in the last 6 years, according to a July 2021 policy report (Grantham Research Institute LSE, 2021). While many of these are aimed at governments, a growing number of cases are targeting a wider variety of fossil fuel and cement companies, but also banks, pension funds, asset managers, insurers, and major retailers, among others. The cases relate to deliberate disinformation (or greenwashing), failure to disclose and manage climate change risk, and some seeking the recognition of corporate human rights. The authors of the report state that "businesses need to be aware of litigation risk," and cite "value chain litigation" as one of three areas to watch in the future. So be warned—the actions and decisions made by procurement professionals, along with those of their colleagues along the value chain, are under growing scrutiny.

## Investors Beware

The authors of the report mentioned above state that in particular "it is likely that more litigation will be brought against financial market actors." This alone is a compelling reason why ESG performance has become a key investment criterion. Investors are increasingly applying these non-financial factors as part of their analysis processes to identify material risks and growth opportunities (CFA, 2020). ESG metrics are not commonly part of mandatory financial reporting though companies are increasingly making disclosures in their annual report or in standalone sustainability reports.

### Sustainability in Action

> In the end, it's a team sport, and everyone is responsible to some extent. We use different approaches to sustainability in our product portfolio. For example, we have sustainability investment products. To a certain extent we can control where funds flow through a process of supply and demand such as in companies that are subject to certain standards and thus promote sustainability.
>
> We also have financing that's linked to the sustainability KPIs of the companies. In other words, the more sustainable the company's development, the more favorable their credit rating.

**Nils Eichberger, Vice President of Procurement, LBBW (Landesbank Baden-Württemberg)**

LBBW is a Mittelstand-minded universal bank with deep roots in Baden-Württemberg; a partner for medium-sized companies, listed groups, institutional customers, savings banks, and retail customers.

## Capital Markets Reflect Corporate Sustainability Trends

Numerous institutions, such as the Sustainability Accounting Standards Board (SASB), the Global Reporting Initiative (GRI), and the Task Force on Climate-related Financial Disclosures (TCFD), are working to form standards and define materiality to facilitate incorporation of these factors into the investment process. Investors, regulators, and stock exchanges are key audiences of sustainability reporting as they play a vital role in using reported information to offer responsible investments and create transparent and sustainable capital markets.

The findings of the seventh annual sustainability survey, published in July 2021 by the World Federation of Exchanges (WFE), were another indication of how sustainability criteria are serious investment considerations. The WFE is a global industry group for exchanges and central counterparty clearing houses (CCPs) that are home to 47,919 listed companies that had a total market capitalization of over $109 trillion at the end of 2020. In essence, the survey showed continued increases in investor engagement in sustainability in both developed and emerging markets, with more than half (50.8%) of the exchanges incorporating initiatives that correspond to all the WFEs five Sustainability Principles—a 10 percentage point increase from the previous year (WFE, 2021).

## The Five World Federation of Exchanges (WFE) Sustainability Principles

In October 2018, the WFE published a set of five sustainability principles ("the Principles") that constitute a formal declaration by the WFE and its membership to take on a leadership role in promoting the sustainable finance agenda. According to the WFE, "the Principles clearly demonstrate that WFE members—while recognizing they are at different stages of market development, with different opportunities and constraints—will seek to promote the Principles progressively in accordance with their circumstances and priorities."

## Employee Expectations: Do the Right Thing

Recent studies have shown that younger generations are more drawn to companies with a purpose, where they can feel a sense of belonging and have an impact that goes contributes to the company agenda and to the greater good. Employers are keen to satisfy these expectations and attract the right talents by offering meaningful work environments and embedding relevant and purposeful actions into their strategies.

## Sustainable Business Cannot Be Ignored

The messages here are clear: sustainability has to be taken seriously and the race to 2030 is on. Companies that have a poor track record of sustainability may find that potential customers will look elsewhere, investors will withdraw their funds, employees will move on, and the authorities may give them some unwanted and uncomfortable attention. Many companies in wide-ranging industries have already understood the urgency to transform and are benefitting from the upsides that are inherent in sustainable business practices.

## A Resilient Future Rests on the Shoulders of Procurement

The long months of suffering under the pandemic brought the supply chain into focus. It changed from being a topic discussed by a select group of supply chain managers and procurement professionals into an everyday talking point. "The supply chain is being discussed not only in the general press, but in virtual social discussions," said an article in leading business magazine Raconteur (Cox, 2020). This is new and with it, supply chain management is quickly evolving to board-level discussions across organizations, which has elevated the importance of the procurement function. In essence, the article highlighted what many procurement professionals have already known for years—that too many companies have underestimated the value of the procurement function, seeing it more as a way of just saving costs.

It took a global pandemic and a worldwide supply chain crisis to finally do it, but this has changed. These recent Earth-shattering events, and before that the financial crisis of 2008, have demonstrated "the dangers of just-in-time strategy and have highlighted the need for one that might better be called just-in-case, which puts much more of an emphasis on resilience" (Ramírez et al., 2020).

The business world is changing, and the accelerated movement toward the "stakeholder economy" means companies are not just accountable to their investors but also to their customers, employees, suppliers, and society at large. More than ever, companies have to pay close attention to the social, economic, and environmental impacts of their actions while at the same time developing greater resilience. To strategically meet these challenges means investing more in procurement, because as recent events have made abundantly clear, supply chain disruptions cause significant losses and immediate impacts to the financials of companies.

## New Purpose for Procurement

We believe in this new purpose for procurement—to create value beyond a sole focus on price by finding the best outcomes together with other stakeholders within the company and externally with suppliers. This value can be added by focusing on resilience, sustainability, and appropriate cost levels. This is the pathway to healthy revenues and profit, improved quality, satisfying consumers' desires, speed to market, innovation, and growth. It is also the pathway to non-financial benefits that will be felt by society at large and the environment in years to come (Fig. 2.1).

This renewed purpose of procurement will need new competencies to challenge the status quo and to strip away unnecessary complexity in the supply chain by working cross-functionally to get the big-picture view of the business, and by building solid relationships with suppliers.

A modern approach to leadership in procurement and new ways of working will empower teams and build a learning and agile organization. Suppliers will be seen as partners.

Procurement will fully embrace digitalization, the enabler of transparency and data-driven insights, which will strengthen decision-making and give procurement teams what they need to focus on those activities that make real, positive impacts on the sustainable future of the business.

Procurement is now on the agenda of enlightened executive officers, and this rise to prominence, combined with its leading role in sustainability, means it will attract the right talent and more resources.

**Fig. 2.1** New purpose of procurement

## Sustainability in Action

> The future supply chain is going to require more partnerships as well as transparency. Data sharing is going to be a key enabler for long-term competitiveness. Hardware and software will both be important. People will have to develop the ability to learn from data to design better in the first place and enhance products out in the field. This will involve the entire supply chain and procurement must orchestrate this transformation.
>
> **Peter Carlsson, Co-founder and Chief Executive Officer, Northvolt**
>
> Northvolt is striving to produce the greenest battery in the world with a minimal carbon footprint and the highest ambitions for recycling to enable the European transition to renewable energy.

This is the time for procurement to shine.

## 2.3 Sustainability Framework for Procurement

> It's about understanding where a company can contribute, which products to focus on, where the biggest footprint is, which technologies can be leveraged, and what business opportunities to focus on
>
> **Sakuto Goda, Principal, Nomura Research Institute (NRI)**

### Sustainability on Top of the Management Agenda

No one these days seriously denies the need for sustainable business practices. Even if they did, the increasing volume of sustainability-related legislation and regulations, consumer demands, and insistence from investors would make it virtually impossible to run a business without at least the minimum of compliance. Sustainability is here to stay—an October 2021 survey found that "92% of companies are adapting their business strategy and models to be more sustainable and plan to increase investments in sustainability initiatives to improve ESG performance in 2022" (Cority, 2021).

This has presented an opportunity for procurement because it has a leading role to play in sustainability transformation. Procurement holds the keys to reducing an organization's footprint as it is the obvious link to the supply chain, which is the source of more than 80% of total emissions (Unknown, 2021). It is procurement that needs to leverage the continuously growing importance of sustainability and to demonstrate that it is the key function to drive positive changes. It is time for procurement to become the driver of sustainability transformation and prompt the re-evaluation of existing processes, influencing product design, proactively suggesting sustainable material substitutes, and to become the catalyst to transform company culture.

Sustainability is not just a bolt-on to procurement's day-to-day operations—it is on a par with quality and cost factors that might normally be key drivers of procurement decisions and should therefore be integrated into existing processes and become part of the working culture. In general, motivating individuals to be mindful of sustainability in their work, and aligning their decisions and behaviors with what is needed to achieve a sustainable future, means engaging and empowering them and other stakeholders from the very beginning. Making this happen involves building awareness and understanding so that all stakeholders are familiar with the purpose, the strategy, and the roles they must play as part of a company's sustainability ambitions.

But how do managers define purpose and strategy? How do they start questioning current processes, and how do they embed a sustainable mindset within the company? Like any other major transformation or change process, a step-by-step approach is strongly recommended, where the overall project is carved up into more manageable chunks. We suggest focusing on five distinct building blocks—purpose, ambition level and strategy, people and transformation, operating model, and material groups. The use of this approach helps to avoid being overwhelmed by the immense scope of a sustainability transformation and helps to ensure that sustainability, bit by bit, becomes an integral part of the business model.

## The Why, What, and How of a Sustainability Transformation

We further categorize these building blocks into the why, the what, and the how of sustainability. The first building block, purpose, equates to *why*, ambition level and strategy are *what*, and people and transformation, material groups, and the operating model are summarized with *how*. We strongly believe that this approach is crucial to ensuring that internal and external stakeholders understand the purpose of the sustainability transformation, what the goals are, and how they will be achieved.

Unfortunately, we have seen quite a lot of companies define sustainability ambitions or targets that they are very unlikely to achieve. The reason for that is simple: compelled to match the competition or to meet customers' expectations, too many companies tend to define market-driven targets and ambition levels without considering their own internal capabilities. Pressures from key stakeholders, including employees, customers, investors, financial institutions, and regulations, can be a positive driving force as long as they do not result in blind actions. Instead, they need to be leveraged to help define the all-important *why* before looking at the *what* and *how* (Sinek, n.d.) (Fig. 2.2).

## Start with Why to Define the Purpose

Asking "why should the company become more sustainable?" helps to clarify thinking from the get-go. For the least ambitious companies, it could simply be a case of complying with regulatory requirements, but this is far from being an inspiring purpose. More inspiring reasons could be to increase market share, gain a competitive advantage, enter new markets, or the altruistic purpose of making the world a better place.

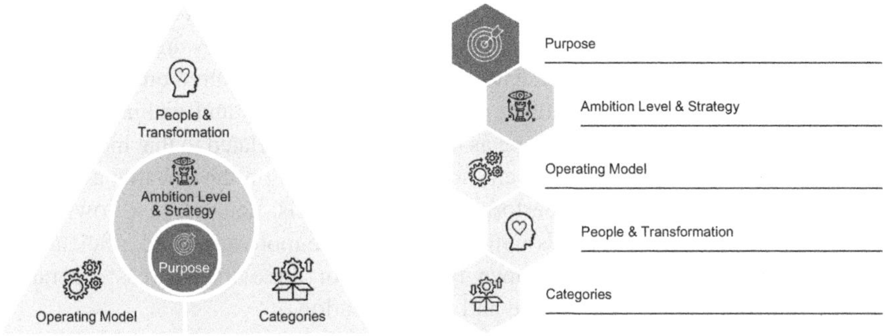


**Fig. 2.2** The sustainable framework for procurement

Asking "why?" is the most powerful way to challenge the status quo and is the foundation of the North Star, a company's unwavering purpose. For this reason, it is important to spend time answering this disarmingly simple but important question. When employees, especially those from generation Z, clearly understand the purpose of the sustainability transformation, how their roles fit into the bigger picture, and how they can contribute, they are far more likely to be motivated to make the transformation a success, especially if they feel that it does some good in the world (Goler et al., 2018).

## Answer the What to Define Ambition Levels and Strategy

Once the purpose has been defined, the next step is to set a realistic ambition level and to translate this into a viable strategy (that should be created to ensure that the North Star will be reached). This step makes it clear what must be done. Is it a case of reducing $CO_2$ emissions? Or does the product or services portfolio have to be changed to make it fully sustainable? Or should there be an overhaul of the supply chain to make sure suppliers comply with social standards and human rights? It is this question that paves the way to specifying concrete measures or activities for team members and certain external stakeholders.

In our experience, a combination of objectives that are ambitious but still realistic with outcomes that are potentially significant is the best way to encourage people to go the extra mile. We strongly recommend defining ambitious but tangible targets that help to successfully reach your North Star.

## End with How to Make Sustainability Tangible

In the end, all strategic planning must translate into tangible actions. It is important that ambition levels are broken down into measurable activities with KPIs and benchmarks.

Sustainability targets in and of themselves may be motivational to employees as a new and exciting device, or even as a point to brag about, but they are mainly drivers of performance—individuals are driven by what they are measured on. Performance can be assured by defining suitable performance indicators and making people or groups accountable for them (as long as the indicators are related to that individual or group's objectives). Performing well, however, depends on the rewards associated with performance. When a performance indicator is defined, and ownership allocated, the associated rewards will foster employee motivation and efficiency.

We suggest defining performance indicators for three focus areas: operating model, material groups, and people and transformation.

## The Operating Model

### Sustainability in Action

A specific procurement team mainly has the responsibility for CSR-related topics, such as ensuring that our suppliers and ourselves remain compliant with regards to working conditions, human rights, and other criteria. We integrate sustainability practices into our overall procurement processes, meaning that each procurement role has responsibilities when it comes to sustainability.

**Henrik Larsen, Chief Procurement Officer, A. P. Møller-Mærsk**

A. P. Møller-Mærsk is an integrated container logistics company and member of the A.P. Møller Group. With a dedicated team of over 80,000 people, they enable global trade for a growing world.

Anchoring sustainability into the operating model is probably the best way to get started. This might involve changing the current organizational setup, revisiting the governance structure, or both. In fact, more and more companies are employing specialists as sustainability managers or as heads of dedicated sustainability teams with the remit to push sustainability within procurement and beyond. Establishing sustainability as part of the operating model means working within existing processes where possible but, for example, building on or adapting them to incorporate sustainability criteria when sourcing decisions are made. Sustainability tools and technology, key enablers of successful transformations, should also be used to create better transparency and to ensure compliance. Making full use of networks and fostering collaboration internally and externally are of utmost importance (Fig. 2.3).

## Categories

### Sustainability in Action

COVID-19 was also an eye-opener for companies that had a high geographic concentration of suppliers. We're now starting to see companies diversifying their supply chain locations and we might even be on the verge of companies starting to in-source vital parts of their production. Even if this is a decision that often means an increase in cost, many companies

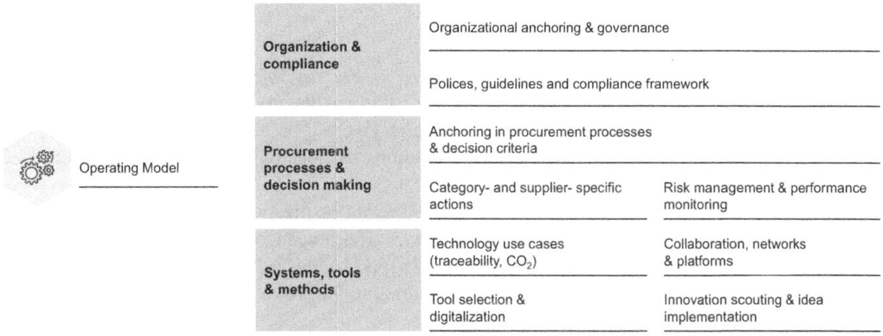


**Fig. 2.3** Operating model

> have realized that the market is more and more ready to pay for the increased certainty and reduced risk.
>
> **Fredrik Gustafsson, Senior Manager of Sustainability Transformation, Cordial AB**
>
> Cordial is a management consulting firm based in Stockholm, Sweden. Its main focus is strategy, transformation, and business development with sustainability as one of the key areas.

This is where procurement is firmly in the driving seat. It is procurement that is responsible for proactively challenging material choices and looking for alternatives when existing materials fail to meet newly defined sustainability guidelines. Performance indicators in this area therefore need to focus on core procurement processes such as supplier or category management.

## People and Transformation

### Sustainability in Action

> Goals and ambitions must be communicated openly, and the vision needs to be clear and understood by everyone.
>
> **Sakuto Goda, Principal, Nomura Research Institute**
>
> Nomura Research Institute (NRI) is a leading consulting firm and systems integrator from Japan. With expertise in the formulation of strategies, globalization of business operations and planning.

People are of paramount importance in any transformation and so, guided by managers, they are the key drivers of change. Managers need to "walk the talk" and be the embodiments of the new strategy, helping to provide support in achieving sustainability ambitions and objectives. They must make sure that teams are engaged and empowered; that they are provided with the right skills, resources, and capacity needed to perform; and that they are surrounded by a fail-fast culture that rewards creativity. Job descriptions and roles may have to change as part of the sustainability transformation, in which case tailored training programs may have to be provided.

## References


Acaroglu, L. (2020). *Decade of disruption: Future of the sustainable workplace: COVID-19, climate change & green recovery*. https://unily.com/insights/guides/future-of-the-sustainable-workplace-in-the-age-of-COVID-19-and-climate-change

BBC. (2021). Climate change: EU to cut CO2 emissions by 55% by 2030. *BBC News*. Retrieved from https://bbc.co.uk/news/world-europe-56828383

Cority. (2021). Sustainability strategy survey finds 92% of organizations increasing investments to improve ESG performance in 2022. *GlobeNewswire*. Retrieved December 3, 2021, from https://globenewswire.com/news-release/2021/10/21/2318321/0/en/Sustainability-Strategy-Survey-Finds-92-of-Organizations-Increasing-Investments-to-Improve-ESG-Performance-in-2022.html

CFA. (2020). *Climate change analysis in the investment process*. CFA Institute.

Cox, J. (2020). *Reshaping and empowering the procurement role*. Raconteur. Retrieved from https://www.raconteur.net/supply-chain/role-procurement-function/

Goler, L., Gale, J., Harrington, B., Grant, A. (2018). The 3 things employees really want: Career, community, cause. *Harvard Business Review* (online edition). Retrieved December 6, 2021, from https://hbr.org/2018/02/people-want-3-things-from-work-but-most-companies-are-built-around-only-one

Grantham Research Institute LSE. (2021). *Global trends in climate change litigation: 2021 snapshot*. LSE Grantham Research Institute Climate Change Laws of the World. Retrieved from https://www.lse.ac.uk/granthaminstitute/climate-change-laws-of-the-world-database/

GRI. (2021). *The Global Reporting Initiative*. Retrieved from The Global Reporting Initiative: https://www.globalreporting.org/about-gri/news-center/gri-welcomes-role-as-co-constructor-of-new-eu-sustainability-reporting-standards

IPSOS Mori. (2020). *Two thirds of Britons believe climate change as serious as coronavirus and majority want climate prioritized in economic recovery*. IPSOS Mori. Retrieved from https://ipsos.com/ipsos-mori/en-uk/two-thirds-britons-believe-climate-change-serious-coronavirus-and-majority-want-climate-prioritised

Kantar. (2020, September). *Who cares, who does? 2020, attitudes and action towards living sustainably and reducing waste*. Kantar.turtl.co. Retrieved September 2021, from https://kantar.turtl.co/story/whocares-who-does-2020-p/page/6

Ramírez, R., McGinley, C., & Churchhouse, S. (2020). Why investing in procurement makes organizations more resilient. *Harvard Business Review*. Retrieved from https://hbr.org/2020/06/why-investing-in-procurement-makes-organizations-more-resilient

Sinek, S. (n.d.). *The golden circle presentation*. Retrieved December 22, 2021, from https://simonsinek.com/commit/the-golden-circle

Unknown. (2021). Using supplier data to measure and manage greenhouse gas impact. *Supply Chain Drive*. Retrieved December 3, 2021, from https://www.supplychaindive.com/spons/using-supplier-data-to-measure-and-manage-greenhouse-gas-impact/592136/

WFE. (2021). *The World Federation of Exchanges Seventh Annual Sustainability Survey shows continued increase in engagement*. WFE Research Team.

WWF. (2021). *Search for sustainable goods grows by 71% as 'eco-wakening' grips the globe*. Retrieved from WWF Press Release https://www.worldwildlife.org/press-releases/search-for-sustainable-goods-grows-by-71-as-eco-wakening-grips-the-globe

# Part II

# From Ambition to Impact

# 3 Develop Purpose, Ambition Level, and Strategy

## 3.1 Defining a Meaningful Purpose

There has to be purpose at the center of a business that needs to be authentic to the company. You can't just copy and paste it from another organization. It is the purpose that gets the hearts and minds of employees, customers, or shareholders involved.


S. Aichbauer et al., *Responsible Procurement*, Management for Professionals,
https://doi.org/10.1007/978-3-030-98640-7_3

**Mark Perera, Chief Executive Officer, Vizibl**

It all starts with creating a purpose. It is a clear, well-defined purpose that gives companies direction and drives them forward, and it is a purpose that guides management and employees. "Starting with why" is the best way to ensure commitment and employee motivation (Sinek, 2009).

It is important for a company to define its purpose and to find good answers to why things are done the way they are. However, purpose should not be confused with vision, mission, or values. In fact, the purpose of a company describes why it exists in the first place, rather than the goals or principles needed to drive that purpose. This is where mission, vision, and values become relevant.

Purpose describes the bigger picture. It is something that is defined to inspire staff and to express an organization's impact on its customers, clients, or investors. In effect, a company's purpose expresses its external impact, or what it does for the people it is trying to serve (Kenny, 2014).

## Having a Purpose Is Key to Success

The COVID-19 pandemic has reinforced the importance of purpose-driven activities. In fact, a 2021 study found that "63% of global consumers prefer to purchase products and services from companies that have a purpose," (Western Governors University, 2021). So, those companies that represent more than what they are selling, that properly communicate their purpose to stakeholders, and that show commitment to their purposes, can influence customers' purchasing decisions. On the other hand, customers may boycott some companies if their purpose is ill defined or unclear in favor of competitors that are purpose driven.

The same is true for employees. Over the past years, as generations Y and Z have taken more prominent roles in the corporate world, employees are actively questioning management decisions and behaviors. They monitor corporate activities, and it is vital for them to be proud of, and to stand behind, what their employers are doing. They have started to actively scrutinize their daily tasks and are craving to see that the work they do has visible and positive impacts on society. If formulated well, a solid purpose can at least partly fulfill these needs and play a key role in keeping employees motivated.

## Sustainability: An Integral Part of Company Purpose

### Sustainability in Action

Sustainability is at the core of our company purpose: we're in business to create a more responsible society by fostering both sustainability and prosperity, which are closely interlinked.

> We see ourselves as change agents for the logistics industry. We've already turned the industry upside down once by providing more transparency through digital solutions—even though people did not clearly express demand for what we're offering. The same thing could and will happen with sustainability.
>
> **Nina Göntgen-Voss, Director of Sustainability, Forto**
>
> Forto delivers a highly transparent, frictionless, and sustainable digital supply chain. Their platform addresses the entire process stream, supporting customers with greater visibility, insight, and control.

It is no secret that there is increasing demand coming from customers, investors, and employees for companies to be sustainable. These demands have been reinforced by the already significant and growing amount of legislation around corporate sustainability. But sustainability does not have to be a burden that is driven by external forces—it can be a powerful lever in developing new business opportunities. Sustainability could be the exactly the right anchor with which companies can formulate their purpose and make it an integral part of their business models.

## Sustainability in Action

> We also use sustainability as an opportunity to expand our business model. For example, we have developed a $CO_2$ dashboard for LBBW based on the demand and consumption information from purchasing, which lowers our $CO_2$ footprint throughout the organization. Every employee, every organizational unit has the opportunity to see and control their emissions. During the expansion, we now distribute the $CO_2$ targets for operational emissions in a balanced score card.
>
> **Nils Eichberger, Vice President of Procurement, LBBW (Landesbank Baden-Württemberg)**
>
> LBBW is a Mittelstand-minded universal bank with deep roots in Baden-Württemberg; a partner for medium-sized companies, listed groups, institutional customers, savings banks, and retail customers.

Before beginning the process of defining a purpose that has sustainability at its core, it is essential to understand just how important sustainability is to the respective stakeholder groups. Sustainability is not new to corporates but since 2020 there has been a marked turning point that has placed sustainability in the spotlight both for consumers and the businesses they buy from. A 2021 study concluded that COVID-19 has led to customers becoming "hyperaware" of the condition of the environment and that they now "seek and champion brands that commit to sustainability" (Forrester, 2021). According to the study, 68% of these "highly empowered" customers will increase their efforts to identify brands that have a positive impact on the environment. But this figure is not only true for customers—employees and investors have similarly high demands. The outcome for companies is clear in that they now have to understand which areas of corporate sustainability carry the most weight and how the essence of the business can help to meet—or exceed—these expectations.

So rather than being a burden, the current emphasis on sustainability has presented companies with the perfect opportunity to take a step back and to rethink why they exist, why they do what they do, and to assess what positive impacts they can have beyond mere profits.

## Purpose-Driven Companies Outperform Their Peers

### Sustainability in Action

One motivating factor is the fear of being left behind. There's a general understanding that sustainability is rising in importance and if you don't act swiftly, it's only a matter of time before you start losing your market share, or regulations catch up and your company becomes obsolete.

Pressure from consumers is also helping to push sustainability in B2C industries. When consumers make up their minds on what's good or bad, companies within that industry need to adapt quickly to stay relevant.

**Fredrik Gustafsson, Senior Manager of Sustainability Transformation, Cordial AB**

Cordial is a management consulting firm based in Stockholm, Sweden. Its main focus is strategy, transformation, and business development with sustainability as one of the key areas.

Companies that operate with a clear purpose also perform better financially. A 2021 report by the Havas Group, a French multinational advertising and public relations company, revealed that brands with an active purpose outperformed the stock market by 206% over the last 10 years. In addition, the report showed that companies with an inspiring purpose also benefit from higher customer loyalty and satisfaction as well as more productive employees who have a stronger sense of belonging (Havas Group, 2021).

Today's companies need an understandable and inspiring purpose if they are to truly stand out and perform well and the same is true for the divisions within an organization, including procurement. Good leaders should define a purpose that inspires their employees and helps them to understand the bigger picture. For procurement professionals, this means thinking beyond price reductions and embracing sustainability. Bearing in mind the current state of our world, there can be no greater purpose than tackling climate change and fighting the injustices too many people are suffering. This is why sustainability should be a priority for everyone who works in procurement.

## How to Define Company Purpose

Defining a purpose is easier said than done. Creating just one inspiring and "on-point" sentence that aims to simultaneously meet the expectations of multiple stakeholders can be an intense and time-consuming process. Flexibility and compromise will be important throughout this process as there will inevitably be the need to accept trade-offs to soothe tensions and disagreements between different stakeholder groups.

But what exactly is needed to formulate an inspiring and on-point purpose statement? Some light can be shed on this by looking at a few prominent examples and trying to understand what they all have in common.

| | |
|---|---|
| Kellogg's | *Nourishing families so they can flourish and thrive.* |
| Coke | *Refresh the world. Make a difference.* |
| Dove | *Discovering the value of 'real' beauty and improving self-esteem worldwide.* |
| Lego | *To inspire and develop the builders of tomorrow.* |

The first thing to notice is that none of these purpose statements mentions an actual product or service. They are all concise, centered around the bigger picture and aim to illustrate the company's positive impact on society—for example, to make sure that families can flourish and thrive, or to improve self-esteem worldwide. Each statement relates to the core activity of what each company does. In the case of Lego, for example, the statement centers on inspiring the builders of tomorrow. Each statement reflects the corporate identity of the respective companies.

However, purpose statements do not necessarily need to be defined on corporate level. In fact, each division in a company could create a purpose statement to motivate employees and to create the required sense of belonging. For this, we suggest following a four-step approach.

**Step 1: Understand What Is Important to Your Stakeholders**

It is vital to understand what is important to the key stakeholders of a division or department and what they value the most. For procurement, the focus should be on elements beyond purchasing. For instance, providing support to make product or service portfolios more sustainable will in turn help to drive sustainability at a corporate level.

Understanding the core values of stakeholders requires an intense self-assessment that involves measuring current impacts and identifying potential pain and gain points. Take a careful look at the daily activities of procurement and where these activities can have the highest impacts in reducing the pains of stakeholders or adding to their gains. Conducting a self-assessment will reveal if the purpose is going in the right direction.

While focusing on key stakeholders, it is important to keep in mind that a purpose statement for procurement has to be in step with the overarching corporate purpose. Any discrepancies between the two should be addressed and the purpose statement for procurement adjusted accordingly.

**Step 2: Focus on Your Key Strengths**

Understand where procurement can have the biggest impacts. What is it that procurement is already doing well and how does this support the company in achieving its overall goals and objectives? What steps are needed to ensure that procurement can create value and facilitate the journey to corporate sustainability?

It is important to think beyond costs and to focus on procurement activities that can drive sustainability, such as product innovation, supplier management, waste reduction, and the use of alternative materials.

**Step 3: Communicate Your Purpose—Internally and Externally**
Once a purpose has been defined, it must be integrated into the core of procurement's activities. This requires clear and constant communication so that everybody understands why it is being done. Try to find alternative and non-conventional ways to communicate the purpose—treating it as an internal marketing campaign could lead to some creative and compelling communication activities that will help to get people to act according to the new purpose.

In addition, make sure elements such as creativity, enthusiasm, and cross-functional collaboration are encouraged when communicating the new purpose. These elements are crucial when it comes to connecting the corporate or divisional purpose with peoples' individual purposes. And keep in mind that good communication is a two-way process—it is not just top-down but also bottom-up, so listen to any feedback and act on it when appropriate.

**Step 4: Act Accordingly and Reassess Regularly**
Defining a purpose in itself alone does not automatically lead to success. Everybody should understand the purpose and act accordingly. Managers should check regularly that their teams truly support the purpose since stringency and compliance with it are crucial for successful operations. Without these checks, there is a risk that the purpose exists as nothing more than a written statement rather than an everyday practice.

In the end, defining a purpose is not a once-in-a-lifetime activity but a series of iterations based on stakeholders' needs. In this respect, it is important to regularly collect feedback from stakeholders to ensure that a particular purpose still fits with their needs. When this is no longer the case, the process of assessing stakeholder needs should be repeated so that a new purpose can be created.

## Purpose and Sustainability Are the Keys to Success

In the end, defining a purpose is the first big step of a sustainability transformation. It is the exercise of defining a Northstar, a communicable goal that inspires and motivates employees and that addresses the needs of key stakeholders.

Defining a purpose is not something that should be done in isolation. It is a process that should incorporate the needs of as many key groups as possible because it will ultimately redefine the relationships between these stakeholders and unify them behind a common sustainability goal. A procurement department, with a carefully considered and properly communicated purpose, can be a powerful force to drive sustainability throughout the rest of the organization. More and more companies are harnessing and benefitting from this force. It is a trend that is set to continue as businesses will pursue purpose and sustainability like never before, mindful this is the best way to retain talent and customers.

"We'll start to see a more purpose-orientated approach in business and in working life," wrote one leading journalist. "The 'great resignation' has encouraged many people to seek out work that feels personally meaningful to them, and I expect this trend will continue in 2022. As employees start to realign their working lives around a sense of personal meaning, it will eventually become clear to employers that purpose is a priority. This will encourage business owners and leaders to take a more intentional approach to strategy and management, figuring out what they can offer their employees and stakeholders, and working to make a meaningful impact in the world" (Skinner, 2022).

There are trends in business just as there are in other areas of modern life, but we believe that purpose linked to sustainability is not just a trend, but a concerted movement for positive and lasting change.

## 3.2 Set Ambition Level

> For the environmental, social and governance parts, our ambition level goes beyond being compliant. It might look like we put a lot of focus on the environmental part since that's what one can read about in the media, but the social and governance parts remain crucial for our way of working. In the industry and regions where we operate these have always been important topics and we'll continue to focus on them as much as the environmental part.
> **Henrik Larsen, Chief Procurement Officer, A. P. Møller-Mærsk**

The journey to become a sustainable procurement organization starts with establishing the company's current situation and ambition level. Before it became the current business zeitgeist, many took the position that sustainability was simply a steppingstone from compliance and risk management and their focus was on external drivers, such as rules and regulations. This is ambition level 1. More recently, many more companies are seeing sustainability for what it truly is—an integrated part of their business that can create real value and a competitive advantage (ambition level 5) (Fig. 3.1).

The decision to go for one of the lower ambition levels could be risky. With rules and regulations changing faster and faster, the risk of being left behind increases when a company sets out to "only" be compliant and neglects to see sustainability as a chance to create a competitive advantage.

### Sustainability in Action

> We keep a close eye on existing and upcoming regulations to be a knowledgeable and helpful partner to our clients. But we won't stop at what is legally required. Our goals and our motivation go way beyond that and double down on our efforts to make supply chains fully sustainable.

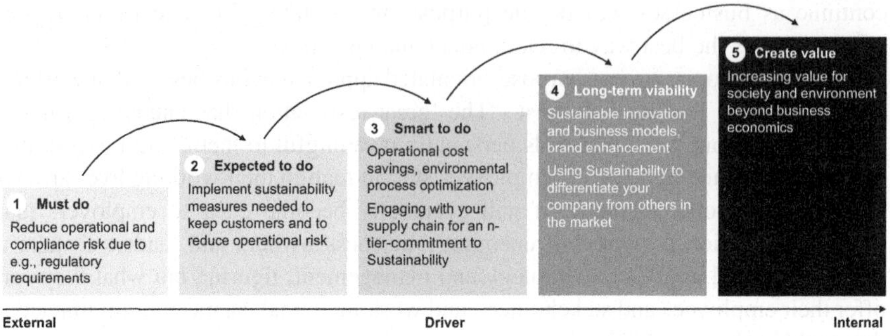


**Fig. 3.1** Define ambition level

**Nina Göntgen-Voss, Director of Sustainability, Forto**

We have also seen that company representatives become very ambitious once they see that the competition has set up their organizations and suppliers to innovate and implement instead of debating about what needs to happen and why. The long-term ambition level should, consequently, be one where value is created beyond business economics. Of course, there has to be technical advancements to support the ambition level, as well as changes to the way all company stakeholders think about sustainability. Both will require a step-by-step approach, which at the very least will get things moving in the right direction. Procrastination will lead, as ever, nowhere new and should be avoided.

### Sustainability in Action

Many companies and countries have ambitious plans and have committed to net zero, but this requires a massive transformation that we're just starting now. When there are lots of companies in this transformation it makes it much easier because a significant volume of emissions come from the supply chain and many companies have shared supply chains. So, a unified movement is a positive start because it will encourage everybody to achieve their targets. The question is: How can you mobilize companies and countries to be ambitious and take action? Tracking progress is now very important because when investors and ESG professionals look at who is performing better we have good metrics to track them. Then competition is on, and this will accelerate the transformation to a zero carbon economy.

**Dr Maria Mendiluce, Chief Executive Officer, We Mean Business Coalition**

We Mean Business Coalition is a global non-profit coalition working with the world's most influential businesses to take action on climate change.

## Two Approaches to Becoming More a More Sustainable Business

As many other companies have already started their sustainability journeys, doing nothing is not an option. The question is, how to begin?

There are two possible starting points. First, company sustainability objectives have already been set, along with specific targets. In this case, procurement can chime in to see how social and environmental issues along the supply chain can be tackled. If this is the case, well done so far. This book provides guidance about how to proceed from this point and the process involved.

Second, sustainability is not yet on the agenda and has to date not been actively pursued by top management, a criterion that is necessary for a successful sustainability transformation. All is not lost if this is the case. Procurement can still take the lead and do all the things within its power to collaborate with team members and suppliers to manage some of the riskiest parts of the business. Again, this book will act as a guide to a journey starting from this point.

Ultimately, whether frontrunner or implementer, the starting point does not matter. Both approaches will drive the required changes. Doing nothing is not an option.

Once the initial hesitation has been overcome and the first successes recorded, working sustainability almost becomes a self-fulfilling prophecy. Depending on how large and influential a company is, there could be political considerations that determine the boundaries that can be pushed by procurement internally and externally. Nevertheless, procurement has a vital role in seeing the bigger picture and advocating change across entire companies and industries; by being bold, thinking big, and joining forces with suppliers and the wider ecosystem, procurement can bundle power and seek real, lasting change on a big scale, and that includes within local authorities, national and international energy providers, and entire countries. It all starts with the right ambition.

## 3.3 Assess Current Situation

Everything stands and falls with transparency.

**Dr Nicholas Garrett, Co-founder and Chief Executive Officer, RCS Global Group**

### The Journey to Sustainable Procurement Starts with Understanding the Current Situation

Taken as a whole, a sustainability transformation journey can seem like a daunting task because this is a decision that affects not just one team or department, but the whole company and eventually its supply chain. After all, a business can only be as sustainable as its suppliers. Whichever way you look at this, it is a huge undertaking, but as with any large project, the key is to take a step-by-step approach. The question now is, where do you begin?

Knowing the starting point is the first step to establishing the route to sustainable procurement and what steps that are necessary along the way. Establishing a baseline involves a combination of internal and external assessments that will provide a clear understanding of the organizational readiness for sustainability, where category-related risks lie, as well as the status of suppliers. Each of these will be looked at in the following sections.

## Organizational Readiness for Sustainability

Several different factors need to be reviewed when assessing organizational readiness for sustainability. We look at this from the standpoint of the operating model—it starts from the strategy and targets, then looks at organization and governance, processes, tools and, potentially *THE* most important aspect to unleash the power from within, the people within procurement and how the culture lends itself to sustainability.

An organizational maturity assessment can be used to get an overview of the current standing and what gaps must be filled before sustainability is truly integrated. To perform such an assessment, we recommend engaging with a range of stakeholders that includes management and members of cross-functional teams, as well as procurement. Keep regional differences in mind, especially in a multinational corporation, where some regions could be more advanced than others from a sustainability perspective.

The assessment can have different levels and starting points, as the following checklist illustrates:

**Strategy and targets**

- Is sustainability part of the company's strategy?
- Is sustainability an integral part of procurement's strategy?
- Is there a company-wide understanding of how to contribute to a (more) sustainable future, and is cross-functional collaboration being nurtured to achieve this objective?
- Have sustainability goals been defined for each element of the sustainability scope?
- Are sustainability goals tracked using defined KPIs, and are deviations adjusted accordingly?

**People and culture**

- Has an internal sustainability culture already been established, and can employees advocate sustainability with suppliers or other stakeholders?
- Has a transformation concept already been established that aligns or actively promotes the procurement organization with regards to sustainable procurement goals?

(continued)

- Are sustainability and sustainability targets part of employee role descriptions?
- Have training activities been planned that aim to develop understanding and skills in relation to sustainable procurement?

**Organization and governance**

- Does procurement have a suitable organizational setup to cover the different aspects of sustainability?
- Does the sustainability department have a strong mandate?
- Have tasks, roles, and responsibilities been clearly defined and assigned?
- Is there a way to monitor changes in the laws and regulations related to sustainability and sustainability trends?
- Have procurement sustainability targets been properly addressed and reflected in cross-functional collaboration?
- Are internationally recognized regulations for sustainable development being observed and complied with?
- Have the minimum legal standards been met and are there aspirational targets for sustainability to go above and beyond the minimum requirement?

**Processes and networks**

- Is sustainability anchored in the category management process and a part of the category strategies?
- Have roadmaps been established with suppliers about how to create the required positive environmental and social impacts, and is this anchored in the collaboration model?
- Have sustainable procurement objectives been clearly described and quantified in the awarding criteria?
- Is sustainability anchored in the risk management process?
- Are sustainability criteria part of the supplier on-boarding process?
- Is procurement, or the company in general, engaged in industry or cross-industry networks and panels?
- Is procurement, or the company in general, engaged in national, international, or supranational councils, panels, and standardization bodies?

**Tools**

- Is there an appropriate IT-system and tools landscape in place to support sustainability and supply chain transparency?

(continued)

- Are third party providers and platforms being used for (risk) control and monitoring?
- Are analytics and predictive use cases in place to support sustainability?

Do not be put off if the answer to many of these questions is "no." On the contrary, it shows where attention is required and gives a true reflection of the current status. Overall, this is an activity that should provide an objective assessment of organizational readiness for sustainability and highlight the focus areas.

## Materiality Assessment: The Key Topics of Sustainability for Your Company

In addition to understanding organizational readiness, it is important to understand how sustainability will impact the overall business. This is where a materiality assessment comes into play. In sustainability reporting, materiality refers to "those topics that have a direct or indirect impact on an organization's ability to create, preserve or erode economic, environmental, and social value for itself, its stakeholders and society at large" (Unknown, 2021).

A materiality assessment is a method of identifying, assessing, and prioritizing the environmental, social, and governance issues that could affect businesses and their stakeholders. The assessment summarizes the issues into a short list of topics that can then be used to guide company strategy, targets, and reporting. Identifying these issues involves reaching out to internal and external stakeholders to get their input, which can be time-consuming, but is also a valuable opportunity to solicit input on strategy.

### Sustainability in Action

We started with a materiality assessment that involved discussions at Group management level from different functions to determine which topics are important to WS Audiology. Once we'd decided on these topics, we then aligned with the team members one level below management and assigned program managers for specific initiatives. The teams below them in turn assigned activity managers who manage part of an initiative. We in the sustainability team are the overall owners of all initiatives—sustainability is one of the three drivers to deliver our ambition to become a clear industry leader.

**Chaojun Li, Global Head of Sustainability, WS Audiology**

WS Audiology develops, manufactures, sells and distributes hearing aids. WS Audiology's ambition as a global leader is to unlock human potential by making wonderful sound part of everyone's life.

A quick look on financial materiality provides a useful comparison to get a better understanding of why a materiality assessment is needed for sustainability. Financial statement items are considered material (large enough to matter) if they could influence the economic decisions of investors. All material matters are to be disclosed. If a public corporation fails to do this, and losses occur, they open themselves to shareholder lawsuits (Banker, 2015). Similarly, those sustainability topics that are considered material could have a major impact on a company and its partners if not dealt with properly.

**The Benefits of a Materiality Assessment**

1. **Engaging in strategy development**: Identifying the company's most pressing sustainability issues is the foundation of creating the right strategy.
2. **Justifying resource allocation**: A methodical and robust process to identify priority issues allows companies to make informed decisions about how to allocate people, budgets and other resources.
3. **Enhancing engagement**: A materiality process requires internal and external stakeholder engagement, so it is a valuable opportunity to gather opinions that can be a springboard for on-going conversations about sustainability. Improved two-way communication with suppliers will lead to a better understanding of the sustainability strategy and improve and strengthen supplier relationships.
4. **Building buy-in**: Involving key stakeholders in the materiality process, and the improved communication that is inherent in that, helps build trust in the outcomes and can lead to better buy-in.
5. **Influencing reporting content and structure**: The materiality process leads to better reporting on the issues that matter most to both the business and external stakeholders (Smith, 2016)

## Assessing Category Risks

Another important step towards establishing a sustainability baseline comes from assessing the different categories and the supply chain. There are a variety of different criteria to consider. Some categories are more significant than others when it comes to sustainability, so it is important to find out where the biggest risks are and where the company is most exposed to these risks.

Bear in mind that categorization based on spend alone is not enough to reveal the full extent of potential risks and where they might lie. There could be multiple categories with a large number of low-value transactions that lie hidden in tail spend that would be considered high risk in the context of sustainability. One approach to this is to use a method such as the sustainability category radar from H&Z to map out category exposure against sustainability impacts (H&Z, 2021) (Fig. 3.2).

The sustainability category radar takes into consideration traditional measures such as spend volume and strategic relevance, as well as the environmental, social, and economic impacts of each category. With access to the right data, supplemented by interviews with suppliers and other stakeholders, it should be possible to conduct this kind of assessment in-house. Alternatively, a dedicated sustainability solution

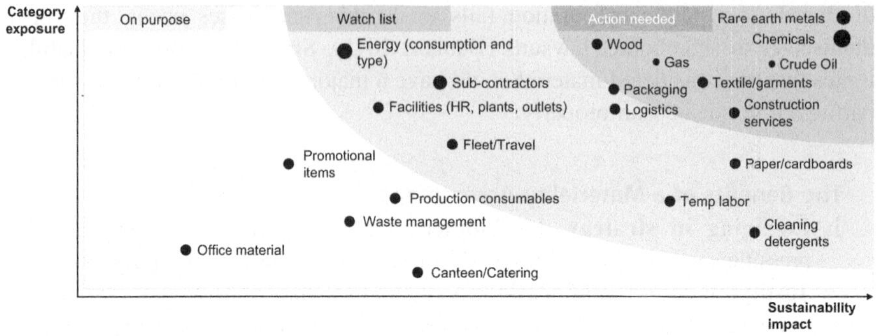


**Fig. 3.2** Sustainability category radar

provider could help by organizing the data for each category and measuring their impacts against the three elements of sustainability.

Indirect categories, such as IT, logistics, packaging, or engineering services, should not be overlooked in these assessments. Even though direct categories will undoubtedly be more prominent on the radar, indirect categories can also be significant. Where direct categories might require individual or industry-specific approaches, the impacts of indirect categories could be assessed by looking at other companies of a similar size.

## How Sustainable Are Your Suppliers?

A company is only as sustainable as its entire supply base. Many global brands have faced some sort of supply chain liability over the years, from coffee to diamonds, and from sports shoes to furniture, and high finance (Kunhardt, 2020). This has heightened awareness of the challenges of tackling sustainability in supply chains, which is good, but many companies have suffered severe blows to their reputations before getting to this point. In the age of social media and a connected world where news spreads like wildfire, the evidence of that reputational cost appears regularly in the news, via intense public scrutiny, and sometimes via consumer boycotts of brands and products. In recent years, some multinational corporations considered to be sustainability leaders came under embarrassing scrutiny because their social and environmental standards were not applied beyond tier 1 suppliers. Apple, Dell, and HP came under fire for sourcing electronics from overseas companies where employees worked in hazardous conditions, and both Nike and Adidas suffered blows to their reputations for using suppliers that were dumping toxins into rivers in China (Villena & Gioia, 2020).

Just as is the case for categories, the supply base also has to be differentiated. The category radar is helpful in identifying those categories most urgently in need of attention, and in turn points towards sources in the supply base that should be prioritized for action.

Even though the category radar will give a first hint about which parts of the supply base should be prioritized, they still need to be reviewed through the lens of sustainability to ensure the right actions can be applied where they are needed. This is an exercise that many companies may already have done already as part of their general supplier relationship management (SRM) practices. However, in this case it must be done with sustainability in mind rather than spend or strategic preferences.

The segmentation can be based on the outcomes of category radar and materiality assessments together with each supplier's strategic importance and potential. The classification needs to be agreed with the relevant cross-functional stakeholders to ensure buy-in from the organization.

The tier 1 supply base should be looked at first, but the lower-tier suppliers are almost always at a higher risk of increasing exposure to potentially serious risks.

## Sustainability in Action

What you don't get on the Internet is how does the midstream actually function. How do European traders, Japanese cathode producers or Chinese refiners link together, and how does the midstream construct link to treatment units and different types of mines in the Democratic Republic of Congo or other high-risk jurisdictions like Indonesia? These parts of the supply chain are more complex and comprise many smaller players that handle far bigger volumes than a Tesla or a BMW would do by securing part of its supply through a direct deal with a mine. At this point there was a massive transparency gap and that's why you needed a more forensic process to gain visibility and to create engagement levers.

**Dr Nicholas Garrett, Chief Executive Officer, RCS Global Group**

RCS Global Group is a proven leader in data driven ESG performance, responsible sourcing, and responsible mining assurance.

It is therefore vital that procurement teams also work towards creating more visibility into the extended supply chain, where 50–90% of environmental, social, and governance (ESG) impacts are found (Jacques, 2020). Consequently, taking a good look at lower-tier suppliers should be a priority as this will provide an understanding of the sub-supply situation all the way down the chain. A digital tool can be of great help once there is this level of transparency, and there are many providers out there in the market that offer a variety of solutions for increasing transparency and tracking sustainability performance.

## Sustainability in Action

Reputation is absolutely key, and the other factor was to offer companies really clear and constructive solutions to dealing with these issues. We break it down for our customers and make an extremely complex set of issues accessible and understandable, we tackle certain elements of it at a time and we isolate certain problems and design out specific processes for these problems or challenges.

We built a unique traceability audit methodology that we still apply widely today which ultimately enables companies to gain visibility over the entirety of their supplier ecosystem, from the auto manufacturer to the mine or recycler. We feed the data into our responsible sourcing technology platform, Vine, which translates the data into dynamic supply chain maps, which include risk scores that are linked to risk data sets derived from our audit process and other sources. The output is "risk heat maps" of companies' entire supply ecosystems, disaggregated by sub-supplier, with a performance score for each supplier. This transparent overview of supplier linkages and performance ultimately enables corporates to conduct risk based due diligence and engage in a risk-based analysis and development of solutions. It's important to take a risk-based approach, as different risk exposure and performance scores call for a variety of approaches to effect continuous improvement. This includes focusing initial attention and effort, often in partnership with the tier 1 supplier, on the worst performers, which can, for example, be 50 sub-suppliers out of 300 identified sub-suppliers in a particular raw material. This ability to manage risk ecosystem wide is new to many downstream companies, as historically their risk management efforts was more directly targeted on their tier 1 suppliers. With our help, improvement can now be achieved with structured corrective action plans also at the tier 4 or tier 5 levels of the supplier ecosystem and we've received a lot of appreciation from companies for this leap in practice. Why? Because the integrity of the entire supplier chain is at risk, when the issues at those tiers of the supplier ecosystem are not adequately addressed and you cannot address them adequately if you only engage your tier 1 supplier.

**Dr Nicholas Garrett, Co-founder and Chief Executive Officer, RCS Global Group**

RCS Global Group is a proven leader in data driven ESG performance, responsible sourcing, and responsible mining assurance.

## 3.4 Develop Targets, Strategy, and Roadmap

Most recently, we have a much more systematic and strategic approach to embedding sustainability in our procurement strategy. The main driver for that is Roche's 10-year ambition with regards to human rights. A 10-year ambition is a powerful anchor. We have to do things differently. It has to be about driving outcomes and actions. Small bets—start small and show the successes. Create some excitement and grow the community.

**Marielle Beyer, Head of Global Procurement, Roche**

### Strategy and Objectives

The information gained from the analysis of the current situation has revealed the organizational readiness for sustainability, what the category risks are, and the state of sustainability among suppliers. In other words, a baseline has now been established, which means a realistic sustainability ambition can be set. The baseline and ambition levels are used to develop or modify the sustainability strategy and to set targets, including financial metrics. However, there is an important interim step that needs to be taken before moving on to the strategy.

### Define the Scope of Sustainability

Sustainability covers a very broad range of environmental, social, and governance topics and not all sustainability-related topics are relevant and applicable to all

**Fig. 3.3** Define sustainability scope

companies. It is important to narrow the focus from the beginning and assess which topics are appropriate for a procurement strategy that has the right balance of environmental, social, and governance targets. The illustration below shows the most important sustainability topics to be considered from a procurement perspective. Narrowing the scope of potential topics will ensure that there is a clear understanding of what is meant by sustainability, something that is vital when discussing it with employees and suppliers (Fig. 3.3).

## Sustainability in Action

At INNIO, sustainability covers all parts of ESG, including the full scope of the GRI standards and the 17 Sustainable Development Goals. We also conducted a materiality assessment that helped us focus on four pillars.

Products—All engines should be ready to be powered by 100% climate-neutral gases by 2025, they should not emit any climate-relevant methane by 2030, and 100% of the materials used to manufacture the engines should be recycled, renewable or reclaimed.

Environment—Greenhouse gases are the focus here since other environmental aspects are already well established. The target is to have production sites reduce their GHG emissions by 50% by 2030 and we want our suppliers and distributors to commit to the same targets.

Employees—We want diversity to increase by 25% by 2025, diversity and inclusion training should have been held for at least 70% of management annually, and employee engagement should reach 85% by 2025.

Society—Here we'll continue to support social initiatives and actively engage with our local communities.

**Thomas Janvier, Vice President of Procurement, INNIO**

INNIO is a leading provider of renewable gas, natural gas, and hydrogen-based solutions and services for power generation and gas compression at or near the point of use.

## Set Long-Term Targets and Work on the Strategy

With clarity and understanding of what sustainability is, and with a realistic ambition in mind, it is time to set some long-term targets. We recommend defining qualitative statements covering all ESG dimensions of what should be achieved first. Is it a case of reducing $CO_2$ emissions? Or does the product or services portfolio have to be changed to make it fully sustainable? Or should there be an overhaul of the supply chain to make sure suppliers comply with social standards and human rights? Questions like these pave the way to specifying concrete targets, measures, and activities for team members. The examples below provide some initial ideas for a straightforward description of target pictures (Figs. 3.4 and 3.5).

Another industry example for sustainability targets focuses on three main areas:

- Optimize supply chain activities according to clearly defined sustainability criteria and encourage suppliers to do the same.
- Create transparency in the supply chain about potential sustainability risks and assess supplier performance.

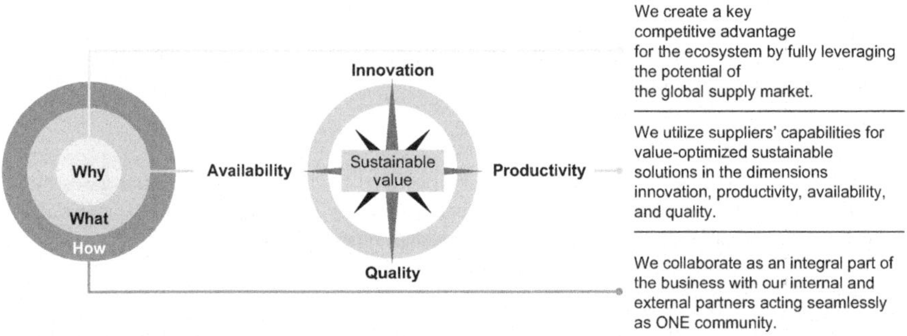


**Fig. 3.4** Industry example for key strategy statements

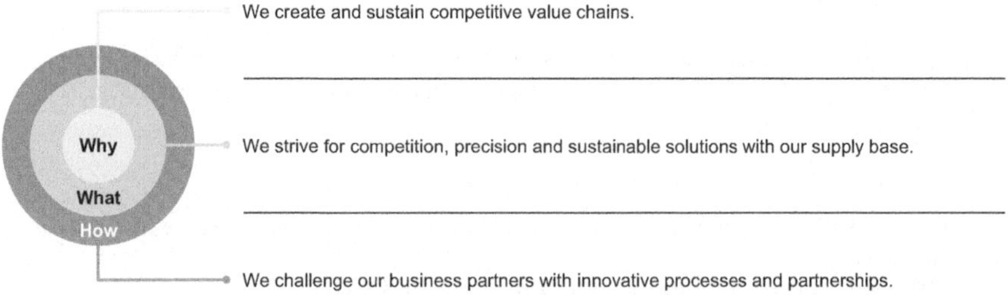


**Fig. 3.5** Industry example for key strategy statements

- Foster collaboration across the supply chain and work together on improving sustainability.

**Sustainability in Action**

We don't yet have specific procurement targets for sustainability but we're working on that at the moment as part of our revised strategy. What we do know is that we want to have procurement-specific improvement measurements that can show our progress.

**Henrik Larsen, Chief Procurement Officer, A. P. Møller-Mærsk**

A. P. Møller-Mærsk is an integrated container logistics company and member of the A.P. Møller Group. With a dedicated team of over 80,000 people, they enable global trade for a growing world.

## Develop Actions and Measures to Deliver on the Targets

Once targets have been defined and described, the next step is to develop the objectives and measures to deliver on the targets. Using the same example as before, the next figure shows how concrete objectives can be added by using specific activities and measurements. It is important to add an (ambitious) timeline to develop the implementation roadmap (Fig. 3.6).

## Align Measures with SDGs

Matching activities to well-accepted frameworks like the UN Sustainable Development Goals (SDGs) is another important step in developing a sound sustainability strategy. The SDG framework is commonly used by companies with an established sustainability strategy because it addresses the full scope of environmental, economic, and social criteria. The strength of this framework lies in its global acceptance and popularity. Using this framework for communications with employees and suppliers helps them to put sustainability into a clear context and therefore leads to better understanding and support for a procurement sustainability strategy.

**Sustainability in Action**

We're focusing our attention where we can make a real impact. We base our reporting on GRI and SASB standards, which give us a very detailed reporting method for how we influence the company and society. While SDGs are designed at national level, an individual company can make a meaningful contribution to some of them.

**Chaojun Li, Global Head of Sustainability, WS Audiology**

WS Audiology develops, manufactures, sells, and distributes hearing aids. WS Audiology's ambition as a global leader is to unlock human potential by making wonderful sound part of everyone's life.

**Supply Chain Sustainability**

| Targets | Target specification | Objectives |
|---|---|---|
| Sustainability criteria inclusion | We optimize our supply chain activities according to Sustainability criteria and encourage our suppliers to follow | ▪ Develop and integrate a supplier code of conduct by 12/2021<br>▪ Develop and integrate sustainability criteria in the supplier processes by 2022<br>▪ Reduce emissions caused by logistics & other Scope 3 activities by 25% until 2026<br>▪ Inform suppliers and encourage a dialogue about sustainability criteria inclusion |
| Supply Chain transparency | We create transparency in our supply chain regarding potential sustainability risks and assess our supplier performance | ▪ Evaluate SC sustainability risks of Tier 1 suppliers and beyond by 2022<br>▪ Assess the risky/main suppliers based on defined sustainability criteria by 2022<br>▪ Audit risky/main suppliers within paper and plastics by 2023<br>▪ Initiate pilot to trace one critical raw material to the source by 2022 |
| Stakeholder collaboration | We foster collaboration across the supply chain to work together on improving sustainability | ▪ Initiate stakeholder dialogue to jointly develop “green products”<br>▪ Engage in associations and industry initiatives e.g., Renewable Carbon Initiative<br>▪ Build network ecosystem to include universities, research institutes, innovative start-ups etc. |

**Fig. 3.6** Industry example for Supply Chain Sustainability targets

## Set KPIs to Measure Progress and Success

The next step involves selecting the right metrics to keep the sustainability strategy on track and then including these into existing targets. Each target and action item will need a concrete and measurable key performance indicator (KPI) against which progress, and success can be tracked and measured. This is often tricky because there is a lack of data and data transparency with respect to sustainability and yet it is this data that is vitally important. That adds another work package to the sustainability roadmap: creating transparency and a data baseline. In the absence of data, work with placeholders until concrete figures can be added once proper data becomes available.

Here are some examples of sustainability KPIs frequently used in procurement:

- The share of recyclable content in materials/products
- Weight reduction targets for purchased products
- Improvement of material-specific properties
- $CO_2$ footprint and scope three emissions in the supply chain
- Total GHG/$CO_2$ emissions from strategic suppliers
- Supplier sustainability scoring (minimum score or targeted average score of the supplier base)
- Spend volume coverage or percentage of suppliers that have signed a supplier code of conduct
- The number of sustainability audits conducted or the percentage of suppliers with a sustainability audit
- The number of corrective action plans with suppliers
- The number of joint supplier collaboration activities

### Sustainability in Action

We also have targets related to the overall inclusiveness and resilience of our supply chain, specifically how well we manage potential risks. We measure this by both the coverage rate but also on the ratings and improvement in ratings.

**Thomas Udesen, Chief Procurement Officer, Bayer, and Co-founder of The Sustainable Procurement Pledge (SPP)**

Bayer is a global enterprise with core competencies in the Life Science fields of health care and agriculture. SPP brings together Procurement professionals to ensure responsible supply chains.

Often procurement organizations measure too many KPIs. It is better to concentrate KPIs around the strategic areas of the company's sustainability strategy. The sustainability KPIs should cover two areas:

- The impact that procurement is trying to achieve, such as reducing GHG and $CO_2$ emissions.
- And the tasks procurement undertakes to bring about that impact, such as a supplier code of conduct, audit coverage or collaboration activities.

The KPIs should then be cascaded down to the relevant teams. Including sustainability KPIs in procurement team targets will help to embed them into everyday practices and processes. But be beware of internal ambiguity or very tough targets as a focus on sustainability and a need for operational savings, for example, might not always go hand-in-hand. Consider this example. A procurement organization has a savings target of 100 million euros. However, adding sustainability targets to the overall strategy, such as reducing GHG emissions, might create conflicts. This means that the savings calculation and TCO methods may need to be re-designed to match this new reality and to ensure that different targets do not contradict each other. In these ways, KPIs also help to communicate the impacts and the trade-offs that need to be made when making critical sourcing decisions.

## Derive a Roadmap

The final step is to create a roadmap that visualizes the strategic plan and the implementation sequence based on the objectives. A sustainability roadmap is generally a 3–5-year plan to elevate procurement to its new strategic status. The roadmap shows how long it will take to achieve the desired objectives and what is involved in developing responsible and sustainable sourcing. By giving a good overview of what is involved, a high-level roadmap also helps to communicate the plan. Nevertheless, a roadmap needs to be accompanied by a more detailed plan that shows the different measures and steps, as well as milestones. In some cases, it might be very difficult to make accurate and confident judgment calls on a timeline and its milestones, but that should not be an issue—getting started is more important than delivering the perfect plan. Timelines can be adjusted at a later date once approval to move forward has been granted and the first steps have been taken. The roadmap should also cover the resources that will be required to achieve the objectives. This includes internal resources, external resources, and tools and technology investments (Fig. 3.7).

Once the roadmap is ready, organize a procurement strategy day to review targets, actions, and implementation steps. This will give the procurement community the opportunity to become familiar with the sustainability strategy and targets. This sort of team review and discussion could also provide valuable feedback for further fine-tuning of the strategy and roadmap.

## Seek Support for Implementation

One of the most common challenges in developing a sustainability strategy is getting buy-in and support from other functions within the organization. Without this buy-in, the strategy is unlikely to succeed. Therefore, it is important that the sustainability strategy for procurement is reviewed both bottom-up and top-down. Company management needs to agree with the procurement team to get the

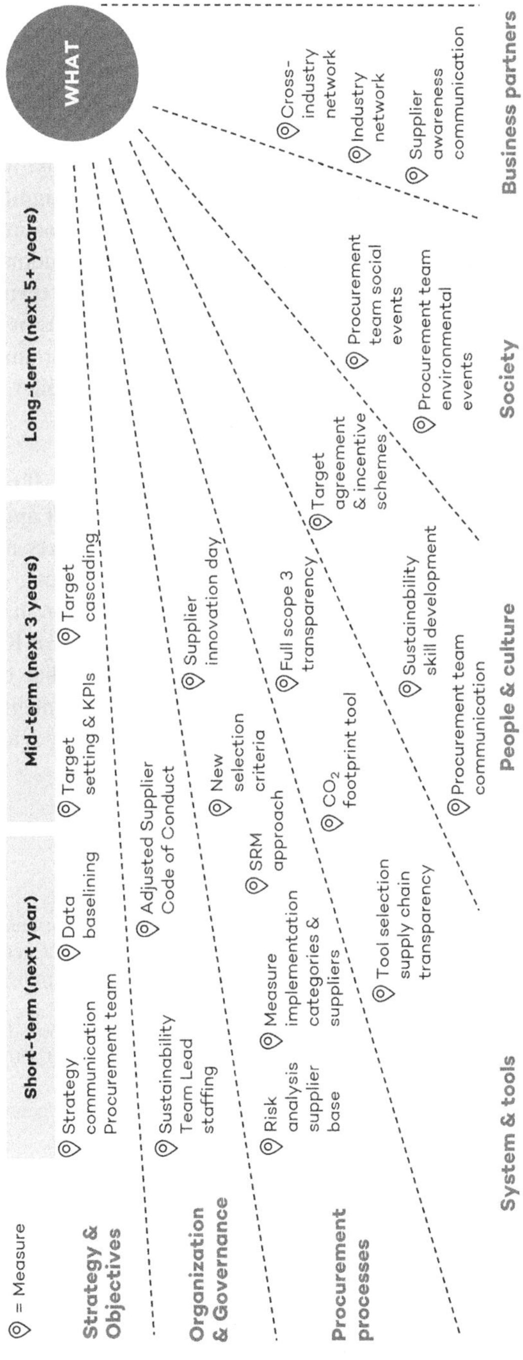


**Fig. 3.7** Example for high-level implementation roadmap

necessary buy-in and resources that will be needed to execute the strategy. This also, importantly, is a step to aligning with other company functions. This is why we see a clear benefit in engaging colleagues from cross-functional teams—such as a corporate engineering, sustainability department, compliance, and risk management—and aligning with what they are doing. Misaligned targets are one of the biggest headaches for internal collaboration.

It is important to move sustainability from a peripheral activity, mostly aimed at managing risks and protecting the company's reputation, into a significant source of innovation, productivity, market differentiation, and growth. Procurement can lead the way on this by developing a strong, workable strategy and roadmap, that weaves sustainability into its operational processes and the mindset of its own teams, as well as encouraging suppliers to do the same thing. This will help inform the company's product and service offering, for example, how its products are produced, marketed, sold, and delivered and to what extent they are reused or recycled.

**Develop Purpose, Ambition Level And Strategy: Summary**
Starting with a new purpose for procurement sets the North Star. If people ask for a purpose, give them one, and let them know the what the level of ambition is. Once both of these are in place, develop a strategy with clear targets and subsequent objectives and measures that will deliver the targets. Matching activities to a common sustainability framework, such as the SDGs, has for many companies helped in developing a sound sustainability strategy. A good set of KPIs are needed to track progress and communicate achievements. Finally, rather than creating a separate document for a procurement sustainability strategy, integrate it into the overall procurement strategy. An ambitious sustainability strategy is a key pillar for a future-proof procurement organization.

## References


Banker, S. (2015). What do supply chain professionals need to know about sustainability reporting? *Forbes Magazine*. Retrieved August 17, 2021, from http://forbes.com/sites/stevebanker/2015/11/14/what-is-new-in-sustainability-reporting-what-do-supply-chain-professionals-need-to-know-about-it/?sh=48b0c1842f40()

Forrester. (2021). Empowered consumers call for sustainability transformation. *Forbes Magazine*. Retrieved December 30, 2021, from http://forbes.com/sites/forrester/2021/01/21/empowered-consumers-call-for-sustainability-transformation/?sh=eb566782042f

H&Z. (2021). *Sustainability as a strategic imperative*. H&Z. http://huz.de/en/consulting/functional-competence/sustainability/

Havas Group. (2021). *Havas' meaningful brands report 2021 finds we are entering the age of cynicism*. Retrieved December 31 , 2021, from http://havasgroup.com/press_release/havas-meaningful-brands-report-2021-finds-we-are-entering-the-age-of-cynicism/

Jacques, R. (2020). *Leveraging technology to optimise supply chain visibility, supply shift*. Retrieved August 4 , 2021, from http://supplyshift.net/resource/leveraging-technology-to-optimize-supply-chain-sustainability/

Kenny, G. (2014). Your company's purpose is not its vision, mission, or values. *The Harvard Business Review*. Retrieved December 29, 2021, from hbr.org/2014/09/your-companys-purpose-is-not-its-vision-mission-or-values

Kunhardt, B. (2020). The knotty challenge of supply chain sustainability (and how to unravel it). *Supply Chain Digital*. Retrieved August 18, 2021, from http://supplychaindigital.com/logistics-1/knotty-challenge-supply-chain-sustainability-and-how-unravel-it

Sinek, S. (2009). *Start with why*. Retrieved December 29, 2021, from http://simonsinek.com/product/start-with-why/

Skinner, E. (2022). Boris Johnson quits as world runs out of Greek letters? Our supporters' predictions for 2022. *The Guardian*. Retrieved January 2, 2022, from theguardian.com/world/2022/jan/01/boris-johnson-quits-global-unrest-supporters-predictions-2022

Smith, M. E. (2016). *Materiality: Five benefits of a great assessment*. Corporate Citizenship.

Unknown. (2021). *Official definition of materiality according to the global reporting initiative*. Retrieved August 4, 2021, from http://youmatter.world/en/definition/materiality-assessment-definition/

Villena , V. H., & Gioia, D. A. (2020, April). A more sustainable supply chain. *The Harvard Business Review* (online edition). Retrieved August 4, 2021, from hbr.org/2020/03/a-more-sustainable-supply-chain

Western Governors University. (2021). *How to develop your company's purpose statement*. Retrieved December 30, 2021, from wgu.edu/blog/how-develop-company-purpose-statement2108.html#close

# Anchor Sustainability into the Operating Model

# 4

## 4.1 Design Organization and Governance

> Many companies have an idea of what they want to do but are missing the pieces in their operating model to truly integrate sustainability. In addition, sustainability makes innovative, new business models possible, especially in combination with digitalization.
>
> **Fredrik Gustafsson, Senior Manager of Sustainability Transformation, Cordial AB**



S. Aichbauer et al., *Responsible Procurement*, Management for Professionals,
https://doi.org/10.1007/978-3-030-98640-7_4

## Organization and Compliance

A significant element of an operating model deals with organizational considerations and governance.

While in the past, sustainability was not even represented as a separate department, or even as a dedicated role, it is now typically embedded in proximity to the CEO and covers the required reporting activities.

When it comes to procurement, many companies are concerned that they need to build up large departments to drive sustainability activities, monitor suppliers, create supply chain transparency, and ensure suppliers' contributions to carbon emission targets, but this is not necessarily the case. At the very beginning, joint-development of processes and experience exchanges of sustainability-related working practices across functions can be a good way to get the ball rolling. This approach has the added benefit of creating an internal network of subject-matter experts in different departments and ensures that the operating model in one function matches the company-wide setup. Establishing a working group can also help to ensure cross-functional buy-in, can be an effective means of implementing a sustainability strategy, and is a step to ensuring that targets are aligned. However, after this initial stage, responsibilities need to be spread across the procurement department, and especially the category managers, to integrate sustainability into all internal and external processes.

### Sustainability in Action

Integration! Sustainability should be integrated into the operating model rather than something that comes 'on top,' which means that all departments need to be closely involved and take ownership of the various sustainability programs.

**Chaojun Li, Global Head of Sustainability, WS Audiology**

WS Audiology develops, manufactures, sells, and distributes hearing aids. WS Audiology's ambition as a global leader is to unlock human potential by making wonderful sound part of everyone's life.

## Sustainability Roles within Procurement

As with all other tasks, it makes sense to create clear roles and define responsibilities for sustainability within procurement. Again, this does not necessarily mean hiring new people, but utilizing existing people who have the right expertise (and ideally enthusiasm for sustainability) to support their colleagues. What is commonly seen is that these *sustainability champions* are part of procurement's processes, methods, and tools team (or similar). The *sustainability champions* play major roles in developing a sustainable procurement strategy that balances environmental protection, social progress, and economic development along the supply chain with the needs of the company.

### Sustainability in Action

There is a procurement team that mainly has the responsibility for CSR-related topics, such as ensuring that our suppliers and ourselves remain compliant with regards to working conditions, human rights, and other criteria. Company-wide sustainability work is driven by our Group Sustainability team and we're now investigating if a separate procurement sustainability competence center makes sense. In addition, we integrate sustainability practices into our overall procurement processes, meaning that each procurement role has responsibilities when it comes to sustainability.

**Henrik Larsen, Chief Procurement Officer, A. P. Møller-Mærsk**

A. P. Møller-Mærsk is an integrated container logistics company and member of the A.P. Møller Group. With a dedicated team of over 80,000 people, they enable global trade for a growing world.

## Category Managers: The Front Line of Sustainability

It is the category managers who play key roles in making sure a procurement strategy is implemented. With their focus on strategic activities and their close connections with suppliers, category managers need to incorporate sustainability factors into the category strategies that support the overall procurement strategy. It is also category managers who typically own significant parts of supplier management processes and thus manage supplier relationships, something that is vital to ensuring that suppliers fulfill their legal obligations. This relationship is also important when it comes to engaging suppliers in joint sustainability activities.

Clearly, category managers need to be professionally trained so they are equipped to deal with the added responsibilities that sustainability adds to their roles. They do not need to be the experts on reporting standards or material substitutes, but they need to know enough to play an active role in ensuring suppliers' commitments and contributions. It is important for category managers to understand that while cost is always going to be a key factor it should certainly not be the only consideration when deciding between one supplier and another. There are "good" and "bad" costs. A supplier may have an outstanding sustainability record but may charge more than its competitors, making this extra cost in theory a "good" cost. A. P. Møller-Mærsk looks at it from a similar angle. Suppliers are selected based on the trade-off cost versus value. Value in ESG terms is not just the cost benefits associated with being sustainable—for example, the cost savings from efficiency inspired by sustainability—but other aspects too, like the value of circularity or positive societal impacts.

### Sustainability in Action

It all starts with our cross-functional Sustainability Review Board, led by the Group Head of Sustainability, which sets the vision and drives the message. [...] So, there is a top-down effort to create awareness. In addition to these conscious efforts, INNIO is lucky that we generally attract people that are sustainability-conscious and intrinsically motivated to drive sustainability.

The Sustainability Review Board consists of leaders from different functions. I'm part of this panel for procurement and the board reports directly to the board of directors at INNIO. Each cross-functional member of the board cascades its messages and actions to their respective function. This also starts with the Sustainability Review Board. Their responsibility is to make the organization aware of regulations and general trends. In procurement, we arrange monthly training workshops and send out targeted communications to different roles to ensure that everyone has the knowledge they need.

As for other roles, there's a sustainability sourcing leader in the procurement team, [. . .] We're breaking down the actions to the different parts of procurement and ensuring that each person in the team knows what sustainability means to them and how it's integrated within their work. [. . .] When we talk about supplier performance, the supplier quality engineers will be responsible for auditing the suppliers with regards to sustainability data—this comes naturally to them since they already have similar responsibilities for quality.

**Thomas Janvier, Vice President of Procurement, INNIO**

INNIO is a leading provider of renewable gas, natural gas, and hydrogen-based solutions and services for power generation and gas compression at or near the point of use.

A good example of how this works comes from the automotive industry, where there is a dedicated procurement role that deals with sustainability without taking the responsibility away from category managers, while also ensuring close collaboration with the team that oversees overall company sustainability.

## What Are Typical Tasks of the Overall Sustainability Team?

- Leading the work to carbon-neutral supply chains.
- Leading the work to increase circularity in materials.
- Leading the work to ensure that the supply chain is responsible and ethical at all levels and tiers, aiming for full transparency and traceability.
- Measuring, reporting, and providing a holistic view of climate action, circularity, and social responsibility in the supply chain.

## What Are the Tasks of the Sustainability Champions Who Work in Procurement?

- Implement the sustainability strategy in the category teams.
- Make the sustainability agenda part of the sourcing decision-making process.
- Build sustainability solutions into the design and development phases.
- Investigate new ideas in the realms of renewable energy, materials, and responsible business in collaboration with the sustainability team (Fig. 4.1).

## Mandate and Governance Bodies

A basic framework for cross-functional collaboration is key to align and drive defined targets in the individual functions and teams. Most companies have already

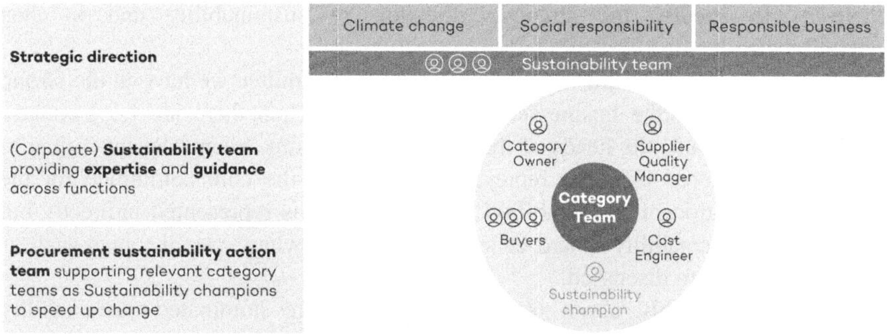


**Fig. 4.1** How to embed sustainability in category teams

created sustainability or CSR boards. With procurement's unique position and relationship, it has with supply chain partners, it should aim for a seat at the table for any company-wide forums where sustainability decisions are made.

While such corporate decision bodies approve strategic directions, investments, and budgets, monitor sustainability progress, and integrate sustainability targets in management compensation and incentive schemes, procurement needs to get clarity regarding its mandate. Typical questions that procurement teams come across include the "weight" of CSR-related topics in supplier selection and evaluation processes. What minimum requirements do suppliers need to fulfill? What are nice-to-haves and what are essential? What are the consequences of supplier non-compliance or a lack of improvement in specific areas?

Procurements need to have the mandate and internal voice to highlight any shortcomings from suppliers and to recommend suppliers based on criteria that stretch beyond savings and price.

## Sustainability in Action

The corporate targets have been given to the units. Supply chain management has responsibility for reducing upstream scope three $CO_2$ emissions. They're responsible for driving this target forward with suppliers and [...] with the individual business units and their procurement departments. Then we focus on changing attitudes to sustainability at corporate level. We've now established internal control requirements for governance where the CEOs are responsible for ensuring that sustainability-related opportunities and risks are implemented strategically and systematically in their business processes and decisions.

We have basically three different blocks. The first one is the governance owners, the next is the businesses, and the last is the countries. Governance owners must define policies, establish management systems, set targets, have KPIs, define long-term goals, and look at budgeting. They are responsible for [sustainability] in their area [...] across businesses and across countries. For sustainability, we have a sustainability board with a sustainability office above it for the group. The task there is to specify the strategic direction of sustainability and to then operationalize it.

Our Sustainability Board is the second-highest committee we have in the Group [...] The CEOs of the businesses are represented. Then there are [...] country representatives [and] the heads of the corporate functions when relevant if they're not already covered by other representatives. [It is] the core committee for the strategic orientation of sustainability[...]Procurement is represented indirectly but attends the Sustainability Board at least once a year when special topics such as decarbonization are discussed.

We have probably about 60-70 people who are nominated sustainability managers in the organization. In the SCM department, there is a separate sustainability unit that takes care of supply chain due diligence law and decarbonization of the supply chain, for example. They are a separate team that defines the policies for the supply chain. They deal with supplier audits and how audits are carried out, what kind of approach is used, and so on.

**Markus Strangmüller, Corporate Development & Sustainability Manager, Siemens.**

Siemens is a technology company focused on industry, infrastructure, mobility, and healthcare, creating technologies and transforming the industries that form the backbone of economies.

## 4.2 Amend Processes and Decision-Making

Companies should include an internal carbon price that reflects its true costs. That will help them to make decisions that are sound. And it provides an internal price signal, so they don't make investments that are carbon intensive.

**Dr Maria Mendiluce, Chief Executive Officer, We Mean Business Coalition**

### Procurement Processes

Sustainability practices and procurement activities are not separate or entirely distinct from each other. Existing process frameworks need adaptions of the core procurement processes rather than the addition of new processes.

This section looks at strategic and tactical procurement as well as supplier and risk management (Fig. 4.2).

### Plan-to-Strategy or Category Management

Sustainability needs to become an integral part of strategic procurement and each category strategy and sustainability levers should be included for each category along the entire value chain. These could include designing-for-sustainability,

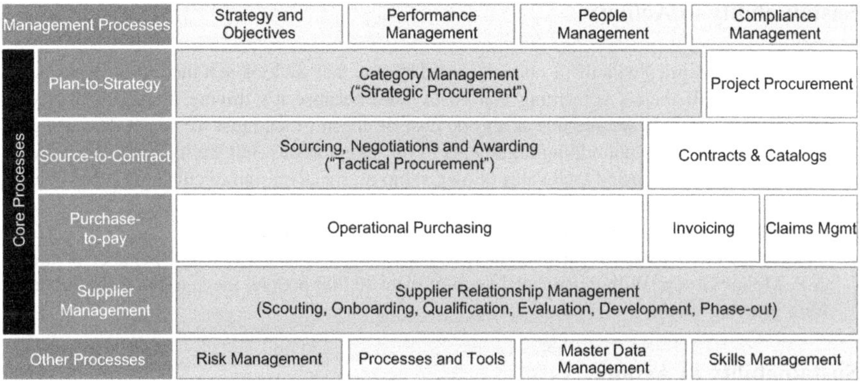


**Fig. 4.2** Procurement process landscape—simplified (H&Z Management Consulting, 2021)

looking at ways to make existing products and services more sustainable, or looking for material substitutes to reduce the sustainability impacts and risks. Done well, meeting category sustainability goals will also deliver better value.

## Source-to-Contract or Sourcing, Negotiation, and Awarding Process

There persist two unfortunate misconceptions in procurement circles that offers with a focus on sustainability are more expensive than their non-sustainable alternatives, and that procurement does not have the mandate to take decisions based on sustainability criteria. These misconceptions need to be corrected. There needs to be a change of mindset so that sustainability becomes one of the decision criteria when selecting suppliers. This means shifting the evaluation models of today to include sustainability topics and to quantify them so that they become part of the total cost. Sustainability should be taken into consideration not only at the beginning of the partnership with new suppliers, but also throughout the relationship. For instance, more and more companies are including sustainability criteria in RFP/RFx/tenders. An increasing number of companies also look at sustainability performance when implementing incentive or recognition programs, thus providing a positive incentive for suppliers to increase investment in sustainability initiatives (Ecovadis, 2021).

Potential suppliers need to be made aware that decisions will be taken based on sustainability at the negotiation and awarding phases. Negotiations are also the right time to look for alternative ideas and solutions that aim at improving sustainability parameters without impacting the price of the product or service. Suppliers should understand that they can be excluded if they do not match the agreed sustainability requirements. Any contract awarded needs to reflect this and include unambiguous, quantified measures and targets that the supplier is expected to fulfill.

## Sustainability in Action

Criteria such as price will always be relevant. The way that we look at it though is good cost and bad cost. If there's something that costs more because it's driving the sustainability agenda forward, then we see this as a good cost. It doesn't mean that we're not discussing prices and that we're not adding arguments to our negotiations, but we look at it from the total cost of ownership and total value of ownership perspectives and calculate what possible positive advantages could come from the extra costs.

**Henrik Larsen, Chief Procurement Officer, A. P. Møller-Mærsk**

A. P. Møller-Mærsk is an integrated container logistics company and member of the A.P. Møller Group. With a dedicated team of over 80,000 people, they enable global trade for a growing world.

## Sustainability in Action

Sustainability is part of the sourcing decision process, and this comes quite naturally since it's also part of our management targets. In many ways the discussion around sustainability follows the same pattern as the one that took place about quality a few decades ago—we're starting to develop more and more mature ways of calculating the total cost of ownership inclusive of sustainability and this will most likely only be stronger in the future. As the cost of $CO_2$ and other sustainability factors increase the business case for sustainable products and services will improve. We don't see a conflict between targets here but rather a complex problem that we're all working together to solve.

Excluding a supplier from the supplier pool for sustainability reasons is something that we can do but prefer not to. When Bayer walks out on a supplier it could mean that the employees of that company end up unemployed and there is nothing that guarantees that the supplier will improve for another customer. That's why we prefer working together with our suppliers to ensure that the requirements are met and that there are continuous improvements.

**Thomas Udesen, Chief Procurement Officer, Bayer, and Co-founder of The Sustainable Procurement Pledge (SPP)**

Bayer is a global enterprise with core competencies in the Life Science fields of health care and agriculture. SPP brings together Procurement professionals to ensure responsible supply chains.

# Supplier Management

When managing suppliers, it is crucial to demonstrate leadership where sustainability is concerned. Companies that push responsibility for sustainability targets further down the supply chain tiers are rarely successful in the long run. In recent years, a rising number of multinational corporations have made it clear that they will only work with suppliers that stick to strict social and environmental standards. Typically, these MNCs expect their first-tier suppliers to comply with those standards, and they ask that those suppliers in turn ask for compliance from their suppliers, and so on, so sustainability standards cascade down the supply chain. It is a good idea, but it has been hard to realize in practice. A more hands-on approach is generally more effective, such as including lower-tier suppliers in the overall sustainability strategy and tasking someone within procurement with extending the firm's sustainability program to first- and lower-tier suppliers (Villena & Gioia, 2021).

Supplier management is even more important when considering that as much as 90% of a company's environmental impacts occur along the supply chain (Quist & Drost, 2021). This is a very high percentage, but on the bright side, it opens up many opportunities for substantial reductions. The process of lowering this number starts at the beginning, by focusing on sustainability during the scouting phase for new suppliers.

In the classification phase, suppliers are prioritized based on a set of criteria. At this point, it is not only the current sustainability status that should be considered, but also the supplier's level of ambition. It can be a struggle to gain visibility into a supplier's performance, but there are an increasing number of digital tools and platforms available that make this process easier (see sect. 4.3 for more details).

The qualification process needs to be adjusted based on the classification. Some industries have started their own certifications for sustainability where individual companies can access and agree to approve suppliers that have been approved by the industry alliance. Together for Sustainability, a global chemical company initiative is a good example. The member companies of this multi-stakeholder initiative have jointly developed key stakeholder requirements, defined a supplier assessment questionnaire, and agreed that any individual member company would accept supplier rating results that came from an assessment by any other company within the alliance.

For high-risk suppliers, a pure tool-based or desktop-like qualification will most likely not be sufficient and in such cases procurement and their cross-functional partners need to form a personal relationship with the supplier and conduct actual visits to verify the situation on site.

Based on the selection and contract performance from the source-to-contract process, the supplier will be measured based on their sustainability performance. The same process is used when it comes to quality issues, for example, when there needs to be a joint effort between the buying company and the supplier to continuously develop and improve sustainability practices.

Procurement and its business partners need to jointly develop clear guidelines about what is tolerable and what is not when it comes to sustainability performance. Typically, a supplier code of conduct describes these expectations. If a supplier does not perform and fails to improve and show ambition, the decision should be made to phase them out and introduce an alternative that does meet the requirements.

**Sustainability in Action**

We've been working with a sustainability agreement for 10-15 years that simply includes certain criteria. We have a two-stage on-boarding process and potential suppliers must satisfy a minimum number of criteria. For example, there's the sustainability agreement that partners have to commit to, which is also audited and, in case of doubt, leads to exclusion if it is violated. This is the basic level, and then we have category specific requirements, which is where we switch from supplier to category management. Individual products then go for certification to Blue Angel, the ecolabel of the Federal Government of

Germany. With consulting or professional services, we expect people to travel by train instead of by car or by plane. This is similar to what we ourselves expect of our employees and where we define requirements for products or services that we purchase.

**Nils Eichberger, Vice President of Procurement, LBBW (Landesbank Baden-Württemberg)**

LBBW is a Mittelstand-minded universal bank with deep roots in Baden-Württemberg; a partner for medium-sized companies, listed groups, institutional customers, savings banks, and retail customers.

Although responsibility for sustainability should, ideally, be shared along the entire supply chain, procurement still has overall responsibility to follow suppliers down to the n-tier to ensure compliance with sustainability standards. This is a major undertaking for procurement, but there are different approaches.

## Innovative Pilot Project Drives Transparency along the Supply Chain

Daimler is increasingly focusing on procurement to clean up its supply chain. Together with Circulor, a start-up specializing in blockchain technology, Daimler is conducting a pilot project for transparency on $CO_2$ emissions in the cobalt supply chain as part of STARTUP AUTOBAHN. The project partners use blockchain technology to track the emissions of climate-relevant gases as well as the amount of secondary material along the complex supply chains of battery cell manufacturers. The data network also documents whether the Daimler sustainability requirements in terms of working conditions, human rights, environmental protection, safety, business ethics, and compliance are passed on throughout the entire supply chain. Daimler calls on its direct suppliers to comply with these standards and requirements, to carry the provisions into upstream value chains, and to monitor their compliance. Through these initiatives, Daimler is driving transparency in the supply chain beyond its direct contractual partners (Daimler, 2021).

## Together for Sustainability: Creating Standards in Chemical Suppliers

At chemical giant Merck, Group Procurement is responsible for integrating sustainability requirements into the relevant stages of its sourcing and supplier management processes. It is a global organization with direct accountability and resources in procurement-relevant local subsidiaries. Its Center of Excellence for Supplier Security coordinates the relevant measures, such as updating its guidelines where necessary, examining processes, and coordinating participation in external initiatives (Merck, 2020). One notable initiative it uses to keep track of its suppliers and ensure compliance with the company's standards and values is its membership of *Together for Sustainability* (TfS) (TfS, 2021), a joint initiative and global network of 34 chemical companies that delivers the de facto global standard for environmental, social, and governance performance of chemical supply chains based on the UN

Global Compact. In addition to conducting its own sustainability audits of selected suppliers, Merck strongly encourages its suppliers to be assessed either on self-reported information or via audits using the TfS framework. This helps member companies to assess the environmental, labor, and human rights, ethical and sustainable procurement performance of their suppliers, and drive and deliver tangible, measurable improvements of their suppliers' as well as their own sustainability performance.

However, it is done, the goal is to create a mindset change among suppliers so that they proactively participate in sustainability measures and develop their own ideas, as opposed to being merely compliant. Just as is the case for discussions about supplier-enabled innovations unrelated to sustainability, this requires relationships based on trust and win–win outcomes. Suppliers need to know that they will gain something by collaboration with buyers to get really good ideas. Generally speaking, the threat of losing business is normally not as effective as positive incentives.

## Sustainability Risk Management

There are clear business benefits to be gained by adopting sustainable practices in procurement but at the same time sustainability introduces additional risks. Sustainability- or ESG-related risks are defined as "uncertain social, governance or environmental events or conditions that, if they occur, can cause significant negative impacts on the company," (Ferma, 2021). Such risks could stem from climate change, water scarcity, disease, poor labor conditions, or new legislation to name a few.

## The Supply Chain Act: Another Valuable Export from Germany?

The latter is particularly pertinent to European supply chain and procurement managers following the introduction in July 2021 of the German Supply Chain Act, the first time that the responsibility of German companies to respect human rights in global supply chains has been given a legal foundation (Soziales, 2021). The Act, which will come into effect in 2023, "places the enterprises covered by the scope of application under an obligation to exercise due regard for human rights and environment-related due diligence obligations in their supply chains. The obligations placed on an enterprise are graded according to its ability to exert influence, specifically in relation to its own business area, the actions of a contractual partner, and the actions of other (indirect) suppliers."

The law means that German companies must either implement new, or adapt existing, risk management strategies to determine whether there is a risk that their own business activities or business activities in the supply chain violate human rights. At the time of writing this affected only German companies, but the EU authorities were paying careful attention with a view to introducing this to the entire bloc.

More generally, from a risk management perspective, accommodating sustainability does not mean re-inventing the wheel—sustainability risk management should use the same methodology as for other risks. It also means that identification and mitigation of sustainability risks along the supply chain are of paramount importance.

## Selecting Suppliers Based on Risk

The size of a supply chain varies from organization to organization and from industry to industry, but it is not uncommon for the supply chain to be vast, with multiple layers of suppliers. From a procurement perspective, it is not feasible to perform a detailed risk assessment for each supplier. Supplier relationships play a key role here, but nevertheless it is also important to prioritize efforts that would be most effective in identifying the suppliers that are exposed to the highest level of sustainability risk.

Typical dimensions to identify potential sustainability risks include category aspects (including the nature of the product, material, or service being purchased) and supplier typology (including risks related to their geo-location, sources of raw materials and labor, historical performance).

### Sustainability in Action

COVID-19 was also an eye-opener for companies that had a high geographic concentration of suppliers. We're now starting to see companies diversifying their supply chain locations and we might even be on the verge of companies starting to in-source vital parts of their production. Even if this is a decision that often means an increase in cost, many companies have realized that the market is more and more ready to pay for the increased certainty and reduced risk.

**Fredrik Gustafsson, Senior Manager of Sustainability Transformation, Cordial AB**

Cordial is a management consulting firm based in Stockholm, Sweden. Its main focus is strategy, transformation, and business development with sustainability as one of the key areas.

## Getting the Right Balance

Managing sustainability risks is a balancing act that relies on developing a greater understanding of the supply chain and how products are produced, where materials come from, and the environmental and social impacts associated with these materials. In other words, assessing and dealing with risk requires greater transparency along the supply chain, something that is greatly assisted using digital technology. This is what the next section looks at.

### Setting Standards

One key point to keep in mind is that it is important not to flood suppliers with sustainability-related requests when there are currently no standards in place. This can make it overwhelming for companies and supply chain participants to use these tools effectively since they may need to fill in sustainability information for several buyers. Each different documentation request could be so time consuming that there is little-or-no time to deal with the sustainability issues.

It would be helpful if tool providers set common standards so information can be shared with suppliers using different tools. Industry collaboration—such as Together for Sustainability—and even cross-industry collaboration, is needed to set these standards and give suppliers a chance to add valuable information without overwhelming them. However, until such standards are in place, digital sustainability tools definitely have a leading part to play in a procurement sustainability strategy.

## 4.3 Leverage Tools and Technologies

> Technology isn't miraculously going to do that for you. You're going to have to tell the technology where to go.
>
> **Dr Nicholas Garrett, Co-founder and Chief Executive Officer, RCS Global Group**

### Sustainability Tools and Platforms

Just as digital tools have sparked a revolution in the way we work and do business, they are also becoming increasingly relevant in tackling sustainability challenges or creating opportunities from them. Digital tools and platforms are key enablers for sustainability. They are being used to collect and analyze data that helps to create greater transparency and shed more light on the state of sustainability within an organization and along its supply chain.

In late 2021, H&Z conducted a survey of more than 140 tool providers to get a deeper understanding of the products they offer, how they approach sustainability, and whether their solutions are designed for widespread use or aimed at specific sectors. The aims were to see what sort of functionalities the tools have and to understand how they can be applied to business functions such as supply chain management, procurement, operations, controlling, manufacturing, or logistics.

**Key Findings: Sustainability Tools as an Enabler of Sustainability—A Market Analysis**

- The market is strong and growing.
- A new wave of young and upcoming sustainability tool providers has entered the market in the past three-to-five years.
- Most solutions cover a broad range of industries and have no specific industry focus.
- The majority of solutions take a holistic approach towards sustainability, covering all three sustainability dimensions—environmental, social, and economic.
- However, a few tools have a clear focus on the environmental dimension of sustainability.

### Sustainability in Action

Digital technology is for sure one of the key enablers for sustainability. However, technology alone isn't able to answer sustainability challenges, as most sustainability topics deal with physical challenges. Think about the concept of the circular economy where products and materials are re-entering the production process through recycling and upcycling with the aim of keeping them within this circle for as long as possible. So, in the end, digital technology isn't the full answer to sustainability challenges. It certainly paves the way as it helps to generate transparency, define your ecological footprint, or to track resources. So, I would say that digital technology should be a pre-requisite as an enabler of sustainability but isn't the silver bullet to solve all the challenges.

**Sakuto Goda, Principal, Nomura Research Institute**

Nomura Research Institute (NRI) is a leading consulting firm and systems integrator from Japan with expertise in the formulation of strategies, globalization of business operations and planning.

## So Many Options Make Choosing Difficult

The sustainability solution market is still in the growth phase. A growing market means lots of choices. However, this could make it difficult to identify the right solution when a particular tool is needed to do a specific job within a certain industry. To make this task easier, the results of this survey were used to categorize the sustainability tool market in terms of the functionalities of the tools, and whether they are niche products or built for broad usage. All solution providers were clustered in a defined structure of categories. These categories have either a company-internal focus or an external focus.

Internally, companies are using digital tools to assess and reduce their environmental footprint, to monitor risks linked to sustainability, and to develop greener production processes. In general, the focus is on enterprise sustainability management, ESG analytics, and carbon and environmental footprint measurement.

Externally, digital sustainability tools are being used to assess environmental, social, and governance (ESG) performance among suppliers, with particular emphasis on supplier transparency, supply chain traceability, and supply chain risk monitoring.

The list below shows the broad categories of tools and what they are used for.

1. *Carbon and environmental footprint*: Measurement of environmental footprint at product or process levels, such as resource usage, carbon emissions, or other greenhouse gas emissions.
2. *Carbon footprint compensation and offsetting*: Reduction of environmental emissions impact by providing solutions in the form of offsetting and compensation projects.
3. *Supplier transparency and traceability:* Creation of product, material, or supplier transparency by aggregating supply chain data to assess the performance for predefined sustainability criteria in the respective dimensions to ensure sustainability standards and requirements.
4. *Supply chain risk monitoring*: Assessment of supply chain sustainability risks based on the integration of various data sources with the objective of identifying risks early and initiating counter measures.
5. *Enterprise sustainability management*: Data integration of, for example, internal processes and operations to assess and analyze internal sustainably performance using defined metrics as the basis to derive improvement measures.

There are many different solutions providers, which, on the one hand, is good, but on the other hand, the sheer number of choices creates confusion when trying to decide on the best option. The technology map below was created to add some clarity by allocating providers to each of the seven categories. The map is published together with Digital Procurement World (dpw.ai), the leading ecosystem for digital procurement (Fig. 4.3).

It is evident that there is no one-stop-shop that covers all functionalities, but most of the solutions are applicable to all industries. The survey highlighted the fact that even though some solutions are well established, there is no single solution that is distinctly different from its competitors and no single solution that technologically stands out from the others.

The survey also revealed some limitations to the solutions. For example, many of them measure $CO_2$ but no other greenhouse gases. In addition, the measurement of emission footprints relies mainly on data from public sources but often the spread is significant as there is no standard procedure in place. This leads to discrepancies and makes comparisons difficult.

**Sustainability in Action**

Digitalization plays a big role in our sustainability work when it comes to retrieving the information that we need from the supply chain. There are a lot of solutions out there and we're currently running several pilots to determine the best way to gather data, how to

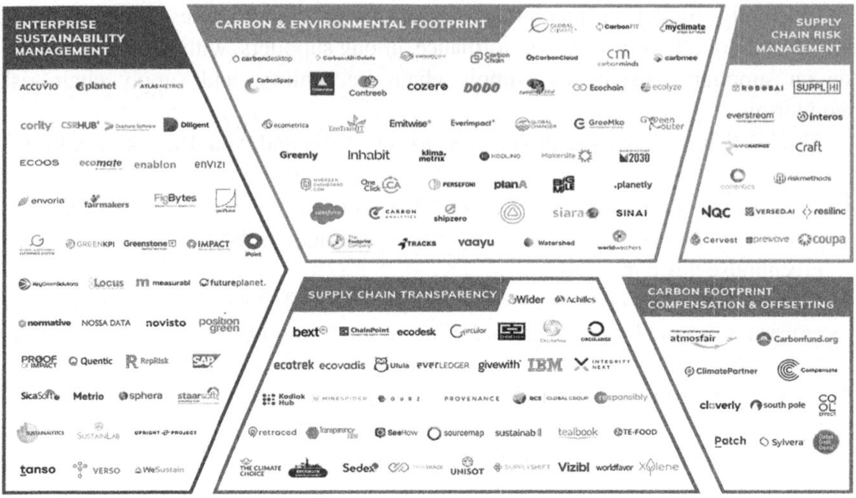


**Fig. 4.3** Technology map (H&Z Management Consulting, DPW, 2021)

analyze it, how to get the transparency that's so important, and then to ensure that everything is properly documented through the entire supply chain and onwards to our customers.

**Henrik Larsen, Chief Procurement Officer, A. P. Møller-Mærsk**

A. P. Møller-Mærsk is an integrated container logistics company and member of the A.P. Møller Group. With a dedicated team of over 80,000 people, they enable global trade for a growing world.

## Practical Implications: Tips on Choosing the Right Tool for the Job

The question is, how do these results and findings help procurement teams select the right tool to match their sustainability goals? There are several practical implications to consider.

- Knowing where to start can be tricky so begin with a risk assessment of internal processes before looking at suppliers' portfolios, including country risks and material categories. It makes sense to start with a tool that increases supply chain transparency around environmental and social aspects as this will highlight where there are issues that need attention.
- Dedicated tools for supply chain transparency or supply chain risk monitoring are very helpful for getting an insight into sustainability issues and risks.
- It is difficult to find one complete solution that covers all sustainability needs. Instead, consider using different solutions that complement each other.
- Identify the key sustainability-related pain points that need to be tackled but keep in mind that there is no single solution that will solve them all. From this point it becomes easier to create a list of key requirements or priorities that the sustainability tool should address.

- Having a list of key requirements, it should be easier to identify the right tool and subsequently build on that.
- Single sign-on solutions are generally preferable, so users do not have to jump from one tool to another.
- Finally, some tools are used by only a small number of users within an organization, but if sustainability is to be a key part of an overall business strategy, they should be accessible by more people.

## Setting Standards

One key point to keep in mind is that it is important not to flood suppliers with sustainability-related requests when there are currently no standards in place. This can make it overwhelming for companies and supply chain participants to use these tools effectively since they may need to fill in sustainability information for several buyers. The different documentation requests could be so time consuming that there is little-or-no time to deal with the actual sustainability issues.

It would be helpful if tool providers set common standards so information can be shared with suppliers using different tools. Industry collaboration, and even cross-industry collaboration, is needed to set these standards and give suppliers a chance to add valuable information without being overwhelmed by multiple requests from different sources.

### Sustainability in Action

> For decarbonization, our initial focus is on the suppliers that we have identified as strategic for Ørsted. For the evaluation, standardized measures are used to make it as easy as possible for the suppliers and avoid it becoming an administrative burden. We make use of the Carbon Disclosure Project (CDP) as a reporting tool. This means that the suppliers only need to answer the standardized questions once and can then share it with as many of their customers as they want.
>
> **Simon Juul Toft, Lead Specialist—Green Procurement, Ørsted**
>
> Ørsted, the world's most sustainable energy company, is recognized as a global leader on climate action and takes tangible action to create a world that runs entirely on green energy.

## A Giant Leap in the Right Direction

Technology alone will not solve all the issues around supply chain sustainability that procurement is faced with, but it will go a long, long way to help. By utilizing the right digital tools or platforms, procurement's understanding of the supply chain and the suppliers in can be taken to a new level. From measuring carbon emissions and footprints, offsetting, risk monitoring, identifying suppliers that need the most attention, or analyzing internal sustainability performance, dedicated sustainability technology is most definitely part of the solution and a vital inclusion in a sustainability strategy.

**Sustainability in Action**

I think, the key difference in how we approached technology was, that we had a laser focus on data just because always, whenever you have sermons about how fantastic technology platforms are, you always have the little caveat at the end which says that obviously the technology is only as good as the data is.

**Dr Nicholas Garrett, Co-founder and Chief Executive Officer, RCS Global group**

RCS Global Group is a proven leader in data-driven ESG performance, responsible sourcing, and responsible mining assurance.

## Practical Example of a Tool Landscape

Digital tools and platforms are an important means of simplifying processes and integrating sustainability into procurement processes. Understanding that not one tool fits all sustainability purposes, it is necessary to combine different solutions to cover the requirements for onboarding and monitoring the (sustainability) performance of suppliers. With most solutions providers offering software-as-a-service or cloud-based tools, integrating the via APIs in core procurement systems is possible. The supplier score can be displayed in the supplier dashboard of the core system, but for more information or deeper analysis a user might need to log-in into the respective tool.

In the example below, an initial screening of suppliers was done via the IntegrityNext platform as a precondition for onboarding. The IntegrityNext platform allows a number of ESG topics to be selected covering all relevant criteria, as well as inclusion and diversity. Existing suppliers are also asked to respond to the pre-built questionnaire. Then, using AI-based screening of news, website, and social media sources, IntegrityNext monitors the target suppliers and provides early warnings and risk alerts if and when necessary (Fig. 4.4).

For a deeper analysis of critical suppliers, the EcoVadis tool is used. EcoVadis allocates scores to suppliers using an evidence-based approach, where suppliers' declarations are credited only when evidence is provided. In addition, ratings are

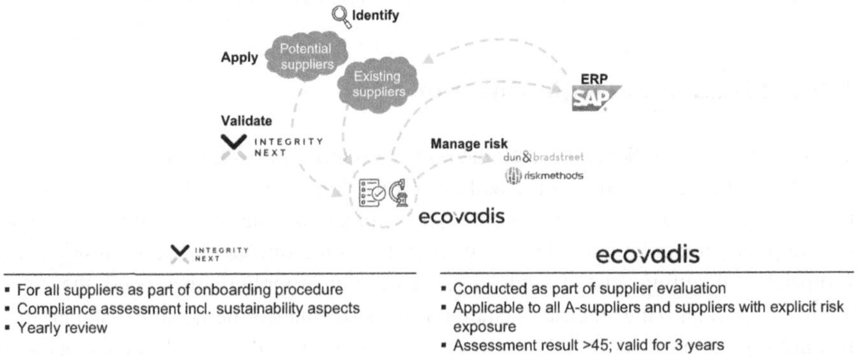


**Fig. 4.4** Example tool landscape

based not only on the supporting documents provided by the suppliers, but also on informed opinions published by NGOs, trade unions, international organizations, local authorities, or other third-party organizations (such as auditors, CDP, or an external compliance database). The EcoVadis score can be regarded as a well-accepted standard.

Additional financial information on suppliers is sourced from platforms available from Dun & Bradstreet, a company that provides commercial data, analytics, and insights for businesses, or similar players such as CreditSafe. And finally, risk management is covered by RiskMethods to provide a comprehensive risk dashboard for the entire supply chain risk management lifecycle.

The example here does not include scope 3 emissions, those produced along the supply chain, but either of the tools mentioned here can support this, or additional tools could be added to the system landscape.

**Ustainability in Action**

I do believe that the other parts of sustainability also get attention, but the climate topic is for sure top of the agenda. Scope 1 and 2 have been measured by many companies for quite some time now but if you want to make a proper assessment of your Scope 3 within the framework of the GHG protocol it becomes much more complex and requires collaboration. Most companies need their suppliers to start the same journey as them, which is a massive undertaking. Since all companies, one way or another, are part of a bigger context in a supply chain, all companies need to make this their focus. Before processes and tools have been fully implemented to measure this, we will most likely continue to see a focus on the environment.

**Fredrik Gustafsson, Senior Manager of Sustainability Transformation, Cordial AB**

Cordial is a management consulting firm based in Stockholm, Sweden. Its main focus is strategy, transformation, and business development with sustainability as one of the key areas.

### It's Time to Make the Smart Choice

Digitalization is an essential component of any transformation, and this is certainly the case with sustainability. The market for sustainability-related digital tools and platforms is thriving and so, with some careful planning and an understanding of what they are needed for, the right tools are readily available.

## 4.4 Enhance Networks and Collaboration

We knew there was a case for building up battery capacities right here ... The European Battery Alliance was created out of this in 2017 and now has 500 member companies.

**Peter Carlsson, Co-founder and Chief Executive Officer, Northvolt**

The challenges posed by sustainability represent a massive undertaking but there is no reason to do it alone. For companies that are starting from scratch, with no in-house expertise linked to sustainability, there are some readily available standards and guidelines produced by NGOs and intergovernmental organizations that would make a good starting point. There are several other types of collaboration that companies can harness to create synergies that could accelerate the transition to sustainable business, including those with industry-specific initiatives, suppliers, employees, and even activist groups. In these collaborations, there is vested interest in sustainability from all concerned so working together results in a win–win.

**Sustainability in Action**

It's about co-creation. Companies need to speak to their suppliers and outside stakeholders to define new ideas and bring them to life. It's about working together to see where sustainability can be leveraged and how new ideas can be developed together. One company on its own will most probably not be able to change society. However, by joining forces and through learning from each other, great ideas can be developed.

**Sakuto Goda, Principal, Nomura Research Institute**

Nomura Research Institute (NRI) is a leading consulting firm and systems integrator from Japan with expertise in the formulation of strategies, globalization of business operations and planning.

## Standards and Guidelines: Pointing Businesses in the Right Direction

Sustainability is more in focus now than it has ever been so there are numerous standards and guidelines, mostly from intergovernmental institutions, that can prove valuable in providing a general overview of what is needed and advice on how to get there. For companies starting their sustainability transformation from scratch, they can be a good place to start. For instance, the OECD's Due Diligence Guidance for Responsible Business Conduct provides practical support, additional explanations, tips, and illustrative examples that "can help enterprises avoid and address adverse impacts related to workers, human rights, the environment, bribery, consumers and corporate governance that may be associated with their operations, supply chains and other business relationships," (OECD, 2018). Similarly, the United Nations Alliance for Sustainable Fashion is an initiative of various United Nations agencies and allied organizations designed to support coordination between UN bodies working in fashion and promoting projects and policies that ensure that the fashion value chain contributes to the achievement of the Sustainable Development Goals' targets.

## Alliances Highlighting Overarching Trends

Several industries already have alliances or initiatives in place in which they collaborate on the main aspects of sustainability that relate to their specific industries. The information ranges from generic to very detailed. For example, the Responsible Minerals' Initiative "provides companies with tools and resources to make sourcing decisions that improve regulatory compliance and support responsible sourcing of minerals from conflict-affected and high-risk areas" (RMI, 2022).

Some of these initiatives stretch across different industries and are politically driven to boost the speed and implementation of specific issues that affect all businesses, offering a myriad of information and potential solutions. A good example is Transform to Net Zero, of which A. P. Møller-Mærsk, one of the companies interviewed for this book, is a founding member. This cross-industry initiative was created with the intention of developing and delivering research, guidance, and implementable roadmaps to enable all businesses to achieve net-zero emissions.

## Networking Platforms

The benefits of using networking platforms such as LinkedIn are clear and obvious, and this is exactly why they are widely used. More recently, SUSTAINX, a solution and networking platform solely dedicated to sustainability, was launched by H&Z Management Consulting and IntegrityNext to combine technology and sustainability expertise in support of companies on their transformation journeys. It facilitates cross-industry knowledge exchange and brings together those in need of solutions with those who offer them.

### Sustainability in Action

It's the decade of action. It's time to learn from each other. It's time to leverage technology and make use of the solutions that are already out there. The world is full of great ideas and collaboration is our generation's lever to make the world become a better place. It is time to start sharing knowledge and technologies, and it is time to really start thinking outside the box. For both, collaboration and networking are key. So, let's stop talking about ambition levels and start translating them into actions.

**Sven Steinert, Director, SUSTAINX GmbH**

SUSTAINX is a web-based platform and community to enable companies' sustainability journey. It offers ready to use solutions and facilitates exchange across industries.

## Suppliers: A Valuable Sustainability Resource within Easy Reach

Suppliers can also be a great source of knowledge and a valuable resource. Enterprises large and small have long realized that there are gains to be had from collaboration with suppliers and opportunities to drive synergies. These might come from joint initiatives on R&D, expanding business reach that is not possible alone, or working to optimize costs for the consumer. It is the same when it comes to sustainability. It is not possible to get an overview of a company's sustainability performance without understanding the practices of its suppliers. This is where collaboration with suppliers is key, because as well as bringing third parties and internal sources together, data from suppliers is needed to get a complete understanding of how supplier performance can impact the supply chain. Collaboration is particularly important where high-risk suppliers have been identified and when it comes to industries and regions that are subject to different rules and regulations.

### Sustainability in Action

We're working together with our suppliers to provide them with what we've learned from our sustainability journey. We don't always have 'the golden answer' but where we've gathered knowledge and experience, we share it with our partners in the supply chain so that the whole ecosystem around a certain process can improve. We see that as an important part of our role within the supply chain and it goes both ways—we're also happy to learn from our suppliers about how we can improve our own practices. These learnings don't only come from our suppliers. We learn equally as much from other partners in our ecosystem, such as partners in our procurement network of non-competing industries.

When we started to introduce sustainability in the supply chain 10 years ago it was still new to many suppliers and also to the internal procurement managers. However, with the enforcement of regulations, along with pushes from consumer and investor, sustainable supply chain becomes a must-have for all companies. Moreover, companies start to collaborate with suppliers on sustainability topics to find solutions to tackle pressing issues and go beyond just compliance.

**Chaojun Li, Global Head of Sustainability, WS Audiology**

WS Audiology develops, manufactures, sells, and distributes hearing aids. WS Audiology's ambition as a global leader is to unlock human potential by making wonderful sound part of everyone's life.

## Sustainability Ambassadors Right under your Nose

There are undoubtedly many employees with a passion for sustainability who are just waiting to be officially empowered and unleashed. Such team members can and should be nurtured for the drive, energy, and ideas they can add to a sustainability transformation. They can also be magnets that will attract like-minded people to join the company, something that should not be underestimated given the desperate scramble to find the right talent that many companies face.

## Activist Organizations: The Checks and Balances of Wrong-Doing

Corporations that come under scrutiny from one of numerous environmental or human rights activist organizations can be subjected to rather painful news headlines and worse. However, these organizations should not be dismissed as anti-business adversaries but viewed as knowledgeable protagonists and potentially valuable collaborators when it comes to creating a more sustainable business.

**Sustainability in Action**

Amnesty International and the likes are very good at keeping issues of societal concern in people's minds and we offer the practical solutions needed to support industry tackling the issues raised by the likes of Amnesty.

Huge credit to the advocacy organizations like Amnesty International or Human Rights Watch who help bring issues to the attention of the global consumer and regulators. They also acted very strategically by not purely pointing fingers at corporations, but also encouraging industry to engage. For example, the advocacy groups made it clear that they did not want industry to try and cut out specific geographies out of their supply chains, but rather they wanted industry to engage in processes on the ground and in the supply chain that would yield demonstrable improvement through collaboration and real effort.

**Dr Nicholas Garrett, Co-founder and Chief Executive Officer, RCS Global Group**

RCS is a proven leader in data-driven ESG performance, responsible sourcing, and responsible mining assurance.

## The Icing on the Cake: Bringing Private Businesses and Non-profit Organizations Together to Boost Environmental and Societal Change

It is increasingly evident that collaboration is one of the keys for unlocking sustainability, with leaders from all sectors of society agreeing that solving environmental and social challenges requires unparalleled cooperation. Collaboration between different types of organizations, business and NGO and multi-sector partnerships for sustainability can produce previously unimagined solutions.

A good example of this is the long-term partnership between KLM Royal Dutch Airlines and the World Wildlife Fund for Nature (WWF) Netherlands, which promoted a market for aviation biofuels (Mousavi & Bossink, 2020).

Another successful partnership between Boehringer Ingelheim, a major international pharmaceutical manufacturer, and Ashoka, the world's largest network of social entrepreneurs and ranked as the world's fifth most impactful NGO of the world (NGO advisor 2021), was set up in 2010 to promote innovative solutions to challenging health problems. "By exploring the intersection of social entrepreneurship, employee talent development, and healthcare networks, the teams have co-designed win-win solutions, including increasing individuals' access to healthcare and community health issues, especially in underserved populations," (Boehringer Ingelheim, 2021). Now in its tenth year, this long-lasting partnership between Boehringer Ingelheim and Ashoka continues to create value while helping to create a fairer society. In November 2021, the partners organized a convention, Making More Health Together 2021, hosting 1400 participants online from across the academic, non-profit, industry, and political sectors to find more sustainable solutions for the most pressing healthcare issues of our time (Boehringer Ingelheim, 2021).

## Sustainability in Action

> We believe the companies that will thrive in the long run are those that integrate social and environmental stakes into their core business, and who understand that the best way to do so is through engaging with systems-changing social entrepreneurs.
>
> We've worked and created partnerships with several very well-known companies who understand the value and importance of a culture of changemaking, including IKEA, Boehringer Ingelheim, Capgemini, Chanel, and others.
>
> **Oda Heister, Managing Director, Ashoka Germany**
>
> Ashoka identifies and supports the world's leading social entrepreneurs, learns from their innovations, and mobilizes a global network to build an "everyone a changemaker world".

Successful business and activist NGO partnerships such as these are undoubtedly why a 2018 survey found that 86% of corporate respondents and 88% of NGO respondents see the role of corporate–NGO partnerships growing in importance (Mizar, 2019). More recently, growing demand from investors and consumers for responsible and sustainable products and supply chains, underlined by new regulations, have been pushing the sustainability agenda to new heights to the point where this is no longer an option but a necessity.

The key to these collaborations is that both NGO and corporate further their own goals while jointly tackling some sustainability-related issues.

### Partnering with NGOs: The Four Keys to Success

1. *Embrace difference—carefully.* Recognize and value the differences between your company and the NGO. After all, without those differences, there would be no reason to collaborate. Alternative perspectives can make collaboration uncomfortable at times, but this is exactly where their value lies—each party brings knowledge, expertise, and capabilities that complement the other's.
2. *Assign a champion and give them authority and decision-making ability.* Partnering with an NGO involves approaching business activities in a new way so assign a champion and give them the power to overcome obstacles and get things done. This also signals to the NGO that the company is taking the relationship seriously.
3. *Make NGO engagement part of the core business.* Partnerships should gradually be integrated into the responsibilities of business units and functions, rather than being maintained as a stand-alone enterprise. Partnerships are most successful when each participant leverages its core capabilities and contributes resources based on normal operations.
4. *Plan for the relationship to evolve over time.* External conditions, interests, and positions will change over time. These could mean that the partnership has reached an end, which is fine. Flexibility and responsiveness are key to how the relationship evolves. It makes sense to build this evolution into the initial agreement so that both sides expect that the relationship will require reassessment from time-to-time (Doh, 2020).

Not surprisingly, there are many useful resources to guide you through sustainability standards, sector-specific alliances and initiatives, and networking platforms. See appendix for a comprehensive list of these resources.

### Anchor Sustainability into the Operating Model: Summary

Get clarity in roles and responsibilities but be pragmatic in order to make sustainability a team exercise across functional borders and with your suppliers. Change your decision-making process and argue and decide for the scenario that creates the best value to the company, beyond just a decision based on a price. You are not alone. Use the power of industry groups and set standards. Embed and combine best-of-breed solutions to get the most benefits for an efficient, future-proof process.

## References

Boehringer Ingelheim. (2021). Retrieved Nov 30, 2021, from ashoka.org/en-sg/partner/boehringer-ingelheim

Daimler. (2021). *Blockchain pilot project provides transparency on CO2 emissions*. Retrieved from Daimler daimler.com/sustainability/resources/blockchain-pilot-project-supply-chain.html

Doh, J. (2020). *Partnering with NGOs: The 4 keys to success, network for business sustainability*. Retrieved Aug 18, 2021, from nbs.net/articles/partnering-with-ngos-the-4-keys-to-success

EcoVadis. (2021, September 21). *Why suppliers are key to unlocking sustainable procurement value*. Retrieved from EcoVadis https://resources.ecovadis.com/blog/why-suppliers-are-key-to-unlocking-sustainable-procurement-value

Ferma. (2021). *People, Planet and Performance: The contribution of enterprise risk management to sustainability*. Tervuren: Ferma.eu. Retrieved from https://www.ferma.eu/app/uploads/2021/03/Ferma-sustainability_2021_final.pdf

H&Z Management Consulting, DPW. (2021). *Technology map*. H&Z Management Consulting.

H&Z Management Consulting. (2021). *Procurement process landscape*. H&Z Management Consulting.

Merck. (2020). *Merck Non-financial report 2020. Darmstadt: Merck, Group Corporate Sustainability*. Retrieved from https://www.merckgroup.com/en/sustainability-report/2020/servicepages/downloads/files/business-ethics-merck-sr20.pdf

Mizar, S. P. (2019, April). *Successful corporate-NGO partnerships*. FM Magazine; FM | Financial Management magazine. https://www.fm-magazine.com/issues/2019/apr/successful-corporate-ngo-partnerships.html

Mousavi, S., & Bossink, B. (2020). Corporate-NGO partnership for environmentally sustainable innovation (March 2020): Lessons from a cross-sector collaboration in aviation biofuels. *Environmental Innovation and Societal Transitions, 34*, 80–95.

OECD. (2018). *Due diligence guidance for responsible business conduct*. Retrieved August 18, 2021, from oecd.org/investment/due-diligence-guidance-for-responsible-business-conduct.htm

Quist, Z., & Drost, R. (2021, March 25). *5 steps to get suppliers on board with sustainability*. Retrieved Sep 22, 2021, from Ecochain ecochain.com/knowledge/finding-sustainable-suppliers/

RMI. (2022). Responsiblemineralsinitiative.org. https://www.responsiblemineralsinitiative.org/#:~:text=We%20provide%20companies%20with%20tools,MISSION

Soziales, B. F. (2021, August 18). *Act on corporate due diligence in supply chains*. Bundesministerium für Arbeit und Soziales.

TfS. (2021). *TfS Initiative*. Retrieved from Together for Sustainability (TfS). https://tfs-initiative.com/

Villena, V. H., & Gioia, D. A. (2021, March-April). *Harvard business review*. Retrieved Mar 22, 2021, from A More Sustainable Supply Chain hbr.org/2020/03/a-more-sustainable-supply-chain

# 5 Set Category Work and Suppliers in Motion

S. Aichbauer et al., *Responsible Procurement*, Management for Professionals,
https://doi.org/10.1007/978-3-030-98640-7_5

## 5.1 Design Sustainable Levers

> Criteria such as price will always be relevant. [...] If there's something that costs more because it's driving the sustainability agenda, then we see this as a '*good cost*'. We look at it from a TCO (Total Cost of Ownership) and TVO (Total Value of Ownership) perspectives and calculate what possible positive advantages could come from these extra costs.
> **Henrik Larsen, Chief Procurement Officer, A. P. Møller-Mærsk**

We live in a very wasteful society. The European Union alone produces more than 2.5 billion tons of waste every year generated by households (8.5%), manufacturing (10.3%), mining and quarrying (25.3%), construction (36.4%) as well as other undetermined waste (18.7%). This a staggering amount that is impossible to visualize. It is primary reason why, as part of its Green Deal, the EU wants to see a change in the way the economy operates (European Parliament, 2021).

The traditional, linear economic model that currently dominates is based on a take–make–consume–throw away pattern. It also includes planned obsolescence, with products deliberately designed to have a limited lifespan. Take smartphones as an example. The average user will get rid of and replace an 'old' phone after just 2 years. Now picture an 80 kg pile of discarded electronic products sitting in the middle of a perfectly normal living room—this is how much e-waste the average American family of four throws out *each year* (Leahy, 2017).

This linear economic model relies on large quantities of cheap, easily accessible materials and energy. It is estimated that as a result of improved technology, increasing demand, and decreasing prices, resource extraction has more than tripled since 1970, and global material use could double to 190 billion tons (from 92 billion) by 2060 (UNEP, 2019). As an illustration of how unsustainable this economic model is, the Geneva Environment Network calculates "Earth Overshoot Day" each year, a date that marks when humanity's demand for ecological resources and services in a given year exceeds what Earth can regenerate in that year. In 2021 that date was July 29, meaning that the resources of around 1.7 Earths were needed to cover demand until the end of the year (Geneva Environment Network, 2021).

Waste, resource depletion, pollution, and loss of biodiversity are some of the obvious consequences of this inefficient and exploitative economic system, which is why the European Parliament and other NGOs have called for change.

This change will come in the form of the circular economy where reuse, sharing, repair, refurbishment, re-manufacturing, and recycling help to create a closed-loop system that minimizes the use of natural resources, waste generation, pollution, and carbon emissions. Fortunately, this system is gathering momentum as more and more companies and their customers benefit from its inherent resource efficiencies. The aim would be to get to the point where the Earth has time to regenerate the resources extracted from it, and the term "Earth Overshoot Day" can be dropped from our vocabulary.

## The Road to a Circular Tomorrow

The road to the circular economy will need to bypass exploitative resource use. Getting there will depend on corporates designing systems and business models that maximize the value of natural resources through innovative production techniques and a conscious selection of raw materials where price is not the determining factor. Procurement will need to be in the driving seat for this journey as it is the key corporate function for identifying sustainable material alternatives and maximizing material efficiencies through reuse and recycling.

The circular economy is based on three simple principles (Ellen MacArthur Foundation, 2021) (Fig. 5.1):

1. Eliminate waste and pollution.
2. Circulate products and materials.
3. Regenerate natural systems.

Although the principles are simple, a fully circular economy cannot be solely designed and implemented internally. Instead, companies that want to instigate a circular economy will have to think beyond their own internal processes, such as engineering, procurement, or production, and include the processes used by their external partners.

The end goal of a fully circular economy is a closed loop of activities where efficiencies are maximized, and waste is minimized or avoided altogether. It is a long-term, incremental process that begins with analyzing internal and external processes to get an understand product lifecycles and the associated impacts of each step.

### Sustainability in Action

Many small and medium-sized enterprises that are further away from the end customer in the supply chain haven't yet risen to the challenge and implemented sustainable practices and processes. It's important for us to keep stressing that this isn't something that only larger corporations have to deal with, it needs to come from all tiers of the supply chain and from all companies regardless of size.

**Thomas Udesen, Chief Procurement Officer, Bayer, and Co-founder of The Sustainable Procurement Pledge (SPP)**

Bayer is a global enterprise with core competencies in the Life Science fields of health care and agriculture. SPP brings together Procurement professionals to ensure responsible supply chains.

Once processes and procedures are understood, collaboration between different functions internally and suppliers externally is needed in order to identify where improvements can be made. Developing a circular economy without the support of external stakeholders is simply not possible, so it is of primary importance that they are included in any and all plans. In fact, enhanced collaboration and co-creation

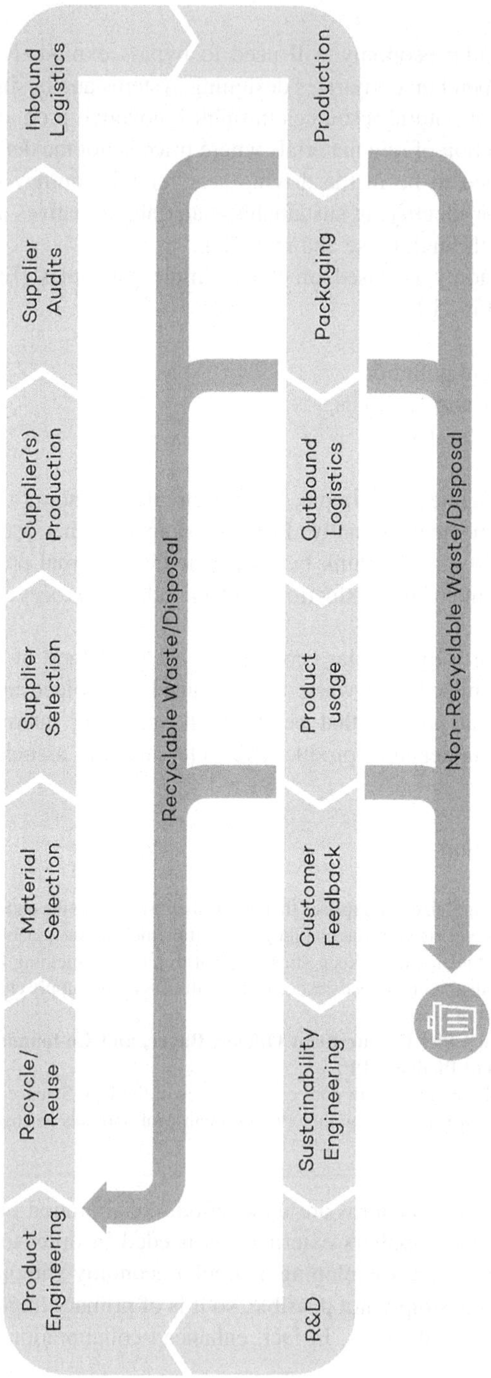


**Fig. 5.1** Circular economy model by H&Z management consulting

with suppliers, as is so often the case, can produce synergies and innovations that would otherwise remain undiscovered.

It stands to reason that procurement must think and act beyond traditional sourcing activities if it is to lead the creation of a circular economy. Modern procurement needs to vertically integrate into the supply chain by pro-actively suggesting new materials to engineering, by scouting the latest technologies, or by collaborating with research and development to bridge any gaps between internal research activities and external know-how from suppliers. In fact, the circular economy will change procurement from a mostly transactional function into a major driver of sustainable transformation.

## Circular Economy: Big Brands Setting an Example for Others to Follow

The number of companies that have publicly announced their commitment to the circular economy revolution is steadily growing and includes some world-leading brands.

Danish shipping giant *Mærsk* has been developing *ways to build recyclable ships* that can be dismantled at the end of their lives and the parts and materials used to build new ships. The company has developed a database called the *"Cradle-to-Cradle Passport" that* documents around 95% of the materials used to build the ships, allowing for more efficient recycling of parts and materials. Instead of mining for new resources, the company will seek to fulfil its needs by using its own products as material banks (Sutherland, 2021).

In the automotive sector, *BMW* launched its RE:BMW Circular Lab in August 2021 (BMW, 2021) while fashion retail giant *H&M* officially launched its innovative circular design tool "Circulator" in November the same year (H&M, 2021). These high-profile companies are not only trying to make their existing products more circular but are aiming to change the way their products are made and how their customers use them.

A common misconception is that it is start-ups that feature most prominently on lists of companies that are leading the way to a circular economy. This is only partly true. Start-ups benefit from one key advantage in that they do not have to adapt existing linear processes but can set up circular systems from day one. But what Mærsk, BMW, and H&M are doing is showing that large, established companies can also be circular economy pioneers by exploiting the two big advantages they have. First, these players have understood the value of the materials involved to such an extent that they decided to take on responsibility for them. And second, they have the business acumen to guide and influence the use phase of a product (Schmid & Ritzrau, 2018).

Other leading companies are now also taking up the challenge.

French energy management and automation company *Schneider Electric* won The Circulars 2019 award in the multinational category due to its efforts to place circularity at the heart of its strategy and innovation. The company uses recycled content and recyclable materials for its products, prolongs the product lifespans through leasing and pay-per-use options, and has introduced take-back schemes to its customers. These activities accounted for 12% of the company's revenues and saved around 120,000 metric tons of primary resources in 2021 (Schneider Electric, 2019).

*Continental*, the German multinational automotive parts manufacturing company, has publicly stated that "by 2050 at the latest, and together with all of its partners in the value chain, Continental wants to make the idea of a circular economy a reality and fully close all product and resource cycles," (Continental, n.d.). Innovative products such Urban Taraxagum™, the world's first bicycle tire made from dandelion-based rubber, and the ContiRe. Tex technology, a yarn from recycled PET bottles, is paving the way to the company's full transition to sustainable raw materials for tire production by 2050.

*IKEA* introduced its buy-back initiative in 2021 in a bid to become a fully circular business by 2030. The initiative encourages customers to sell back their "gently used" furniture to the company in exchange for IKEA store credit. The used products are then resold in the store's AS-IS section (IKEA, n.d.).

*Timberland*, the American manufacturer and retailer of outdoor wear, started its sustainability journey in 2008 by introducing soles on its footwear made with recycled crumb rubber from discarded truck and car tires. It later developed a recycled rubber outsole compound made with at least 35% post-industrial recycled latex, a material that would normally go into landfills. However, its biggest step towards circularity happened in 2014 when the company, together with Singapore-based tire manufacturer Omni United, created shoes that are made from Omni's used passenger vehicle tires. This is a great example of cross-industry collaboration, and one that should serve Timberland well in its bid to make 100% of its products fully circular by 2030.

## Key Levers for a Circular Economy

Despite operating in widely different market sectors, all the companies mentioned above have one thing in common: they plan and organize their activities by using sustainability levers, the so-called Rs of sustainability. These range in scope from the well-known 3-Rs—reduce, reuse, recycle—to a much more extensive list 14 different levers. By analyzing examples from some prominent companies, and combing this with first-hand experience from a variety of consulting projects undertaken by H&Z, we suggest focusing on ten sustainability levers (Table 5.1):

As far as procurement is concerned, the first (smarter use of materials) and third (useful application of materials) of these sustainability levers are the most relevant to the respective sourcing categories. They emphasize the need to identify new technologies, to find sustainable material substitutes, and to ensure that existing suppliers are leveraged as part of the drive towards a circular economy.

The good news is that there are already quite a lot of innovative technologies or sustainable materials out there. However, identifying the best suppliers and

**Table 5.1** Sustainability levers used in a circular economy (adapted from Potting et al., 2017)

| | | |
|---|---|---|
| Smarter use of materials | 1: Refuse | Stop using materials that have a negative impact and find sustainable substitutes or material alternatives. |
| | 2: Rethink | Challenge the status-quo by pro-actively suggesting materials, new technologies, and innovations from suppliers. |
| | 3: Reduce | Minimize the use of natural resources. Try to focus on renewable and biodegradable alternatives. |
| Extend the lifespan of product and its parts | 4: Reuse | Think beyond end-of-life and by designing and developing products and materials that will have a second life—Either as part of another product or in another industry. |
| | 5: Repair | Try to repair and maintain defective products so they can be used with their original functions. |
| | 6: Refurbish | Restore current products and bring them up to date to ensure their functionality matches current trends. |
| | 7: Remanufacture | Use parts or materials from discarded products to produce new products with the same functions. |
| | 8: Repurpose | Use parts or materials from discarded products to produce new products with different functions. |
| Useful application of materials | 9: Recover | Make sure that product parts or materials can re-enter the production cycle by getting them back from customers and then forwarding them to suppliers. |
| | 10: Recycle | Ensure that products and materials are recycled so that they re-enter the production process, or that they can at least be used for other products (low-grade quality materials). |

discovering these innovations remain significant challenges. This is where the concept of "green lever workshops" comes in and where procurement professionals should be making full use of technology.

## Green Lever Workshops

Cross-functional lever workshops are not new to procurement. In fact, they have been held regularly by procurement professionals for a number of years as a method of optimizing external spend. H&Z Management Consulting developed its proprietary Bull's Eye tool to assist with this process. Centered around four perspectives—demand, specification, sourcing, and execution—the tool helps procurement professionals to identify category-specific levers in order to ensure maximum cost efficiency (Fig. 5.2).

With sustainability, there are ten levers to consider rather than four, but the general idea of using the tool to define green levers remains the same:

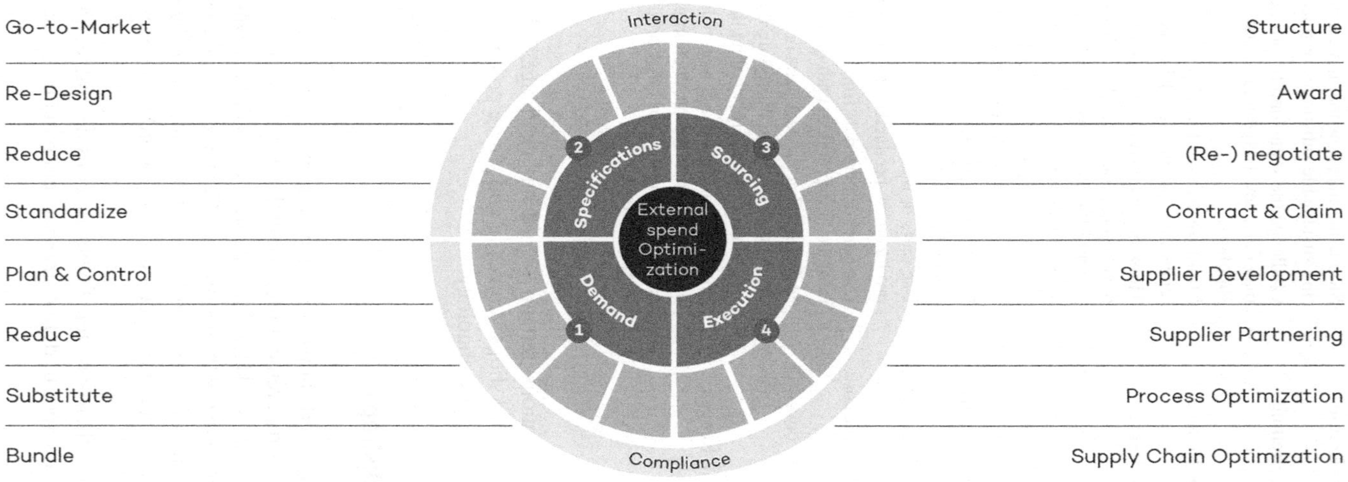


**Fig. 5.2** The H&Z Bull's Eye

1. Identify fields of action.
2. Define key levers.
3. Develop an implementation roadmap.
4. And start discussions with suppliers.

There are four rules that should be considered when going through this process.

**Rule 1: Select the Right Input Sources**

Traditionally, one of the key measurement parameters for procurement is spend data—a status quo that is currently shifting due to sustainability-related activities. With the growing importance of sustainability, prices or spend data are no longer the only sources of truth. Companies now also need to understand the environmental and social impacts of their materials and sourcing categories. In turn, this means that procurement has to understand the value and transportation chains of suppliers.

**Sustainability in Action**

> We categorize suppliers based on spend, category, country, and type of product or service to identify those that are the highest risk. [. . .] In addition to standard audits, we have a process called product-related environmental protection (PREP) to ensure that there are no forbidden materials in the products, and that restricted materials stay within assigned limits. This special auditing process is run directly by the sustainability team.
>
> **Chaojun Li, Global Head of Sustainability, WS Audiology**
>
> WS Audiology develops, manufactures, sells, and distributes hearing aids. WS Audiology's ambition as a global leader is to unlock human potential by making wonderful sound part of everyone's life.

Over the past few years, procurement professionals have increasingly been faced with new questions such as "What is the $CO_2$ footprint of material A?" or "Do our suppliers comply with human rights regulations?" The answers to these questions will only grow in importance. In order to successfully apply green levers and to define sustainable actions, it is therefore important to thoroughly understand the impacts of sourcing categories and to be familiar with category-specific pain points via a materiality assessment.

Apart from changing measurement parameters, it is important that there is a shift in focus from top-tier suppliers to the smaller suppliers within the supplier base. In the past, procurement tended to center its main strategic activities around those suppliers that accounted for 80% of spend volume but with the demands of sustainability, this strategy will have to be re-thought. Modern procurement departments need to take a risk-based approach that assesses the risks of their entire supplier base. This can be based on factors including geographical location, social and environmental performance, or by checking generally accepted certificates. Although this task might seem complex and time-consuming, it can be done relatively easily nowadays by using technology. That said, there is one general rule to keep in mind: smaller suppliers are more likely to carry sustainability risks than their larger peers.

Stakeholder interviews should also be conducted as a way of understanding internal pain points within sourcing categories. It is crucial to collect feedback from internal customers to get a clear picture of what their key requirements are, and which areas need to be emphasized when looking for sustainable alternatives.

**Rule 2: Generate Ideas within Cross-Functional Teams**
Once all relevant data has been collected, it is time to define some concrete actions. This is when collaboration between cross-functional teams is essential, especially where sustainability is concerned. Procurement is the most likely source of innovation, but it alone will not be able to change products or services, which is why other functions such as engineering, research and development, production, or marketing have to be involved to make sure that product portfolios can become circular.

### Sustainability in Action

Collaboration happens precisely at the process interfaces. But it's not a one-way street. We approach the product lifecycle management community and ask them to what extent they consider sustainability-related opportunities and risks in their process. It's not about defining every milestone, it's also about a change of mindset—we want to get our design and product teams actively thinking about sustainability and circularity within the framework of a journey. This includes potentially using different materials and that's where there's strong collaboration with supply chain management.

**Markus Strangmüller, Corporate Development & Sustainability Manager, Siemens**

Siemens is a technology company focused on industry, infrastructure, mobility, and healthcare, creating technologies and transforming the industries that form the backbone of economies.

The strategic buyers also work together with the sustainability manager to develop specific targets for their categories. Within indirects, for example, we identified energy efficiency levels for real estate and leased properties. For direct categories, we will partner with engineering to reach 100% sustainable material inputs. We 're currently setting the baseline for this and we will then agree on measures to take.

**Thomas Janvier, Vice President of Procurement, INNIO**

INNIO is a leading provider of renewable gas, natural gas, and hydrogen-based solutions and services for power generation and gas compression at or near the point of use.

Procurement should identify sustainability solutions for the respective sourcing categories in preparation for these cross-functional workshops. It is advisable to collect a few potential solutions that overcome the challenges and pain points that have been identified and that have the chance of making specific products more sustainable. These solutions will function as a baseline for discussions to see, for example, if engineering would need to change the specifications of a product, to check if a production line would need to be changed to accommodate new material, or to make sure that research and development is familiar with the material changes in case further (company specific) research would be required prior to the implementation.

**Rule 3: Make Full Use of Technology**

It is crucial to leverage technology to analyze supplier risks and to identify sustainable solutions. There is an extensive range of sustainability technology providers that offer innovative solutions to match a wide variety of needs. As there are so many tools available, identifying the most appropriate can be a challenge in itself but it is nevertheless worth the effort to develop the right tool landscape. Section 4.3 gives a good overview of sustainability tools and technologies.

Technology can also help to identify material substitutes or to learn about best practices and innovations. In fact, using technology will significantly reduce the effort needed to stay up to date by continuously screening. This is precisely why the SUSTAINX platform and community was developed. It offers state-of-the-art sustainability solutions and connects professionals with sustainability experts to discuss specific challenges and to co-develop new solutions.

Naturally there will be some challenges that cannot be overcome with existing solutions. However, this is not a reason to be discouraged from getting started. A lack of technology to solve a specific challenge is the perfect opportunity to being collaborating and co-creating with suppliers, or to make use of the expertise available through SUSTAINX working groups.

**Rule 4: Define a Reasonable Implementation Roadmap**

The last crucial step is to define an implementation roadmap beginning with some sustainability quick wins, or solutions that are fairly easy to kick off yet still produce noticeable outcomes. Examples of these quick wins include the creation of a sustainability-related code of conduct, rolling out green IT solutions, or installing devices that remove $CO_2$ from production sites. These quick wins can be very diverse in where and how they make an impact, so it is important to find out early on what the biggest pain points are and the best ways to address them.

**Sustainability in Action**

> Our strategic buyers are the ones integrating sustainability clauses in our supplier contracts. We have some standard language regarding ESG in our general terms of purchase, but we believe in agreeing on individual targets based on what the material aspects of sustainability are for each specific supplier.
>
> **Thomas Janvier, Vice President of Procurement, INNIO**
>
> INNIO is a leading provider of renewable gas, natural gas, and hydrogen-based solutions and services for power generation and gas compression at or near the point of use.

Finally, procurement should strongly consider working with their cross-functional teams on sustainability lighthouse projects. These are projects that remain true to the core business but create public awareness of sustainability or circular economy initiatives. Timberland's recycled rubber outsole or Continental's yarn from recycled PET bottles that were mentioned earlier in the chapter are excellent examples of innovative lighthouse projects that won these companies very positive and highly valuable public recognition.

### Start Thinking in Circles

It is time to re-think the traditional way of doing business. The linear economy is no longer fit for purpose as the potentially catastrophic costs of its take–make–consume–throw away pattern is no longer acceptable or justifiable. Fortunately, many companies are recognizing this unquestionable fact and making the transition to a circular economy. This is not a fad led by start-ups as there are some real business heavyweights that are leading the way and demonstrating the economic sense of circularity while at the same time minimizing the negative impacts of their businesses on the environment. They are doing this by focusing on the three core principles of circularity: eliminate waste and pollution, circulate products and materials, and regenerate natural systems.

Procurement should be one of the key drivers of a circular economy. This can be done by including the 10R green levers framework in discussions with suppliers instead of just focusing on cost efficiencies. This will require a far greater degree of collaboration with cross-functional teams internally and supply chain partners externally, the targeted use of technology, and more willingness to co-create to find solutions to sustainability-related challenges. These actions lay the foundations for a circular economy that will ultimately bring enormous benefits to business, society, and the environment.

## 5.2 Foster Cross-Functional Collaboration: Sustainability Is a Shared Responsibility

> I think that's why we're orchestrators and facilitators of the ecosystem with a cross-functional team.
>
> **Mark Perera, CEO, Vizibl**

### Secure Top-Level Commitment

For a sustainability strategy to be implemented and ultimately successful, it must start at the top. The responsibility for sustainable sourcing, as well as all the other defined actions, cannot rest solely with procurement and other functional business sections. Those at C-level have leading roles to play and they need to understand the impacts and leverage that procurement has.

*CEOs* must see the big picture and the opportunities that can be found in running a sustainable business. Changes in the supply chain can have huge impacts on well-informed consumers, especially Generation Z, and therefore open up new revenue streams. Embedding sustainability and getting a mindset shift across the organization depends to a huge extent on strong, authentic, and inspiring leadership.

*CFOs* need to understand sustainability from a risk perspective. Reputational and legal risks in the supply chain can jeopardize an entire business, even if the goods procured seem to be insignificant in value.

*COOs* must come to grips with how sustainability can influence make-or-buy decisions, and then, together with procurement, develop strategies to minimize the ecological footprints of their decisions, reduce risk exposure, and boost business.

Finally, *CTOs* need to introduce whatever changes are necessary to the products themselves and be allocated the resources to do this.

## Engage Stakeholders and Form Working Groups

On a functional level, or on a day-to-day working level, to be more precise, especially for procurement, it is of key importance to join forces across and beyond functional pillars. Silos have no place in a sustainability strategy. It is important to identify the key stakeholders and learn how best to manage them. It is also important to look at ways of accelerating sustainability efforts and find out what measures and resources are available to do this.

### Sustainability in Action

Active dialogue with our stakeholders is one of the ways, be it regulators, customers, shareholders, employees or communities. We're there to deliver for our stakeholders so if they need a voice, we'll listen. It's the same for procurement, which is the customer of our suppliers, and is therefore responsible for voicing the need for sustainable actions in the supply chain. Stop reading this book and get started today!

**Thomas Janvier, Vice President of Procurement, INNIO**

INNIO is a leading provider of renewable gas, natural gas, and hydrogen-based solutions and services for power generation and gas compression at or near the point of use.

The main responsibility for the sustainability work within Ørsted lies with the Global Sustainability Department. Responsibility for supply chain decarbonization lies with the sustainability, HSE and procurement teams in unison whereas the actual execution of supplier decarbonization lies with HSE and procurement.

**Simon Juul Toft, Lead Specialist—Green Procurement, Ørsted**

Ørsted, the world's most sustainable energy company, is recognized as a global leader on climate action and takes tangible action to create a world that runs entirely on green energy.

Sustainability is a cross-functional task for the entire organization as all functions—procurement, production, engineering, product development, but also corporate strategy—are confronted with sustainability challenges. Making product portfolios more sustainable, developing production processes that are more sustainable, and finding material substitutes or sustainable material alternatives are in the interests of the entire organization. Orchestrating all these activities within the organization is one of the tasks of the sustainability manager but to do this effectively, it is important to enhance cross-functional collaboration and ensure that people talk to each other, that they work with each other, and that they strive towards common sustainability goals. It is for this reason that this is now a vital role and one that will continue to grow in importance especially compared to 5 years ago when sustainability was seen more as a task than a dedicated function.

### Sustainability in Action

My team and I are responsible for driving sustainability within the organization as well as the value chain. We have two focus areas: integration and communication.

Integration—Sustainability should be integrated into the operating model rather than something that comes 'on top,' which means that all departments need to be closely involved and take ownership of the various sustainability programs.

Communication—It is also crucial that we communicate our sustainability efforts and performance to the relevant stakeholders and the general public. We communicate via our sustainability report, our website, social media, and other outlets.

**Chaojun Li, Global Head of Sustainability, WS Audiology**

WS Audiology develops, manufactures, sells and distributes hearing aids. WS Audiology's ambition as a global leader is to unlock human potential by making wonderful sound part of everyone's life.

## Cross-Functional Collaboration Is Key

There are some interesting similarities between the rise to prominence of sustainability on the corporate agenda that we see today and that of digitalization a few years ago. From an organizational point of view, digitalization was the domain of IT departments, and from there separate, small teams were usually spun off to investigate what needed to be done. Today, there are digitalization experts and data scientists spread across organizations who have responsibility for IT landscapes and roadmaps.

In advanced organizations, sustainability is an embedded function with ambassadors and subject-matter experts in all key corporate functions. But nevertheless, the responsibility for networking and making connections does not just fall upon the shoulders of these ambassadors—it is a cross-functional activity that requires involvement, input, motivation, engagement, and action from all company functions. In other words, it is not possible to have too many people involved in sustainability because it is already an important driver behind most functions.

### Sustainability in Action

For me, procurement is one of many roles that plays a part in sustainability. We have two core targets regarding the supply chain: first, an ESG secured supply chain, and second, $CO_2$ reduction within the supply chain. These are essential to our long-term objectives. But decarbonization of our own operations is also important. Overall, we see sustainability as a business opportunity so it's part of our strategy discussions. It's also an important consideration with the PLM or the product designers because they aim to develop products that help society to decarbonize or move into the circular economy. So, procurement has an important role to play in sustainability, but so do many other functions.

Each relevant unit has a sustainability manager. The sustainability managers are then in a community that we orchestrate. There are meetings between the governance owners only, or with only the countries or the businesses. On top of that, there are topic-related, cross-functional sessions. Then, every two years, we usually have a sustainability managers summit.

**Markus Strangmüller, Corporate Development & Sustainability Manager, Siemens**

Siemens is a technology company focused on industry, infrastructure, mobility, and healthcare, creating technologies and transforming the industries that form the backbone of economies.

## Employee Motivation in Procurement Teams and beyond

Many employees will be intrinsically motivated to work on sustainability goals, and some may even become real driving forces behind the work. This is ideal, but even in these circumstances, it is important that the targets are aligned across different functions. If, for example, procurement puts more focus on sustainability, this needs to match the targets of cross-functional partners.

Another point to consider is that putting sustainability higher on the agenda may, in some cases, mean more work for procurement. This needs to be reviewed carefully to ensure that the extra work does not put a strain on the team. Again, it is about mindset change, transformation, and about creating an environment where everybody can speak up and contribute.

### Sustainability in Action

We had a sustainability fair more than two years ago. We managed to share ideas and discuss sustainability with 280 colleagues for a day: We had booths, we involved external partners, and very importantly, we had a series of workshops where colleagues could contribute their own ideas for the further development of ideas related to the planet, credit, carbon footprints, and personnel. This generated a lot of momentum, and we did a bit of advertising about the topics and gained a lot from it. And then at some point it becomes a bit of a self-runner.

**Nils Eichberger, Vice President of Procurement, LBBW (Landesbank Baden-Württemberg)**

LBBW is a Mittelstand-minded universal bank with deep roots in Baden-Württemberg and a partner for medium-sized companies, institutional customers, savings banks, and retail customers.

## 5.3 Unleash the Power of Suppliers and Partners

Traditionally, suppliers have been seen as extensions of companies rather than as integral parts of their core businesses. Flows into plants have usually been optimized using the most advantageous internal costs and timeframes but not necessarily on the flows or cost balances within a global network of multiple suppliers. A lack of perspective and a preference for specific suppliers have tended to diminish the possibilities of new ideas and cost reductions. The desire to develop new products and more variations of them has not helped to keep costs and complexity down, nor have these inventions really created something new and valuable for customers.

Then, during the sourcing process, suppliers have generally been told that they need to come in at their very best prices, no matter how stretched, from the moment they start supplying, in the belief that the procurement team would not ask for more yearly savings than the usual efficiencies that can be created by the supplier—something that never really materializes. Terms such as trust, partnership, as well as tolerance for making mistakes and fixing them together, have not been part of procurement's vocabulary when talking to suppliers. The relationship between procurement teams and their suppliers has been strained, primarily because it has been based on cost only.

Some may regard this as an exaggeration of the traditional relationship with suppliers, or even dismiss it completely. The point is that, generally speaking, businesses could do better when it comes to creating value for themselves, their supply partners, their shareholders, their employees, society at large and the planet. This has never been more relevant and urgent than it is now when the world is facing problems that can only be fixed through a spirit of widespread collaboration rather than competition—or a false sense of security buyers have just because they have multiple suppliers in the price race. As is evident from the numerous examples in this book, those who collaborate can move mountains.

The most successful companies already recognized this quite some time ago and have developed relationships with their current suppliers, potential new partners, and their internal stakeholders that are based on sharing, that are open and honest, so that everyone involved gets to enjoy the business. After all, suppliers are a great source of knowledge when it comes to finding alternative and more sustainable materials and processes; they can be invaluable in paving the way for early discussions about implementing circularity in the design and concept phases; and suppliers can be resourceful where avoiding waste during development, manufacturing, and logistics processes are concerned. In addition, supply chains that were created jointly give both partners responsibility and shared ownership. This brings the benefits of having more eyes to see what might possibly go wrong, more hands to help when disruptions arise, more hearts to inspire and engage teams on the journey, and more brains to figure out at any point in time the optimal mix needed for resilient, cost competitive, and sustainable supply chains.

## Sustainability in Action

If it's not carbon, it's going to concern water, your forestry, or maybe diversity, so you need the data, but you also need the engagement with your suppliers. That's what's fascinating because it's going to change the way that companies work with their suppliers. It's not a dictatorial approach where you say you have to do this. You need to take them on the journey with you as well. It's the ability to align your sustainability goals with your suppliers and then it's the ability for you to collaborate with them around the transformation revealed by the value chain.

**Mark Perera, Chief Executive Officer, Vizibl**

Vizibl is a digital procurement platform that allows global organizations to unlock the true value of their supplier relationships.

The obvious question that arises at this point is exactly how to develop a more open approach to supplier collaboration, especially when sustainability spans so much of the supply chain. In our experience, it takes awareness and transparency, being an ambassador for suppliers, securing action and implementation, pushing relentlessly, and celebrating and repeating.

## Awareness and Transparency

Once procurement has laid out its overall sustainability strategy and objectives, the next step is to explain to suppliers the reasons for the new strategy. When suppliers understand the "why" behind the strategy, it becomes much easier for them to understand the rationale of their customers and subsequently develop ideas about how they can walk this new sustainability-oriented path together. This may seem obvious, but in our experience, it is too often overlooked because people have developed a misguided belief over the years that giving away information is tantamount to giving away leverage. This is an outdated attitude that needs to change.

On top of that, not all suppliers have a clear idea about their own sustainability ambitions and how sustainability can be embedded into their strategies and decision-making processes. In fact, it is somewhat alarming just how many companies want to play it safe where sustainability is concerned and take the minimalist option of complying with rules and regulations only. This is a blinkered approach that will come back to haunt suppliers that take this route—once regulations increase (and they will) and they find that they do not have the right internal structures in place or have not prepared their own organizations for the transformations of their customers, the result will be lost business. Collaboration is particularly important where high-risk areas have been identified and where industries and regions are subject to different rules and regulations.

Compliance to legal standards and regulations is vital, but more commitment is needed to meet the challenges posed by sustainability. This begins by sharing information about the overall strategy and its targets either by inviting suppliers to take part in regular dialogues about sustainability, or by asking them to participate in a larger event (live or online). Getting an interesting keynote speaker to talk about a "hot" sustainability topic, showing some success stories, and inviting suppliers to bring in their ideas can be very motivating. A CEO address is a sure-fire way of demonstrating that sustainability, and the new supplier relationships this involves, is a top priority. These do not have to be large, lavish events. In some cases, it would be more appropriate to tailor a sustainability session by giving it a particular theme such as decarbonization, conflict minerals, workplace diversity, or human rights along the supply chain.

## Being an Ambassador for Suppliers

Close collaboration with suppliers and other stakeholders can be a very effective way of generating new ideas that provide fruitful returns for all concerned. Needless to say, when it comes to figuring out precisely what substances are part of a component, identifying alternative materials, changing designs for a more circular approach, or rerouting logistics routes, for example, it will be necessary to collaborate with the relevant stakeholders in the company so that new ideas will be accepted and ultimately materialize. It is an important ingredient of the success of these collaborations that all relevant peers are involved in discussions from the beginning. Since most organizations work in categories, it is recommended that all necessary partners within the company are engaged in plans to tackle sustainability in specific categories. Keep in mind that some materials might need an approach that goes beyond a specific category. Large-scale challenges such as decarbonization, plastics, and packaging, for example, are going to need careful coordination and someone who can take the lead to push such initiatives through the organization and with suppliers.

### Sustainability in Action

> We're working together with our suppliers to provide them with what we've learned from our sustainability journey. We don't always have 'the golden answer' but where we've gathered knowledge and experience, we share it with our partners in the supply chain so that the whole ecosystem around a certain process can improve. We see that as an important part of our role within the supply chain and it goes both ways—we're also happy to learn from our suppliers about how we can improve our own practices. These learnings don't only come from our suppliers. We learn equally as much from other partners in our ecosystem, such as partners in our procurement network of non-competing industries.
>
> **Henrik Larsen, Chief Procurement Officer, A. P. Møller-Mærsk**
>
> A. P. Møller-Mærsk is an integrated container logistics company and member of the A.P. Møller Group. With a dedicated team of over 80,000 people, they enable global trade for a growing world.

## Securing Action and Implementation

There is no such thing as one-size-fits-all when it comes to creating and implementing a sustainability strategy. There are numerous pragmatic tools that can be used to lay the foundations and cover the basic sustainability requirements, but for higher risk areas, or where there are significant $CO_2$ emissions or human rights violations that must be dealt with, different approaches, or a mix of approaches, should be used.

### Approach 1

*Set up a hackathon:* A hackathon (a combination of the terms "hack" and "marathon") has its origins in software development. Relevant experts work in one room over a day or two and collaborate intensively on specific projects. The goal is to use the time to come up with functional results or as many ideas as possible for pre-defined outcomes. It is an approach that is well suited to sustainability initiatives. It can be an internal company event or even one that includes supply partners or other experts.

### Sustainability in Action

We organized a sustainability fair more than two years ago and managed to share and discuss topics around sustainability with 280 colleagues and external partners for a whole day. An important aspect were the workshops in which people could actively participate and bring in their own ideas to develop solutions along the four lines of planet, credit, carbon footprint and personnel. Of course, this generated a lot of momentum, and we did some advertising to spread the word ... and then, at some point, it became a bit of a self-runner.

**Nils Eichberger, Vice President of Procurement, LBBW (Landesbank Baden-Württemberg)**

LBBW is a Mittelstand-minded universal bank with deep roots in Baden-Württemberg and a partner for medium-sized companies, institutional customers, savings banks, and retail customers.

### Approach 2

*Approach individual suppliers* that create high carbon emissions, or those that generate a lot of plastic, and schedule workshops aimed at finding alternative ways to manufacture the materials and final components. Some suppliers may already have started doing this. If that is the case, join forces with them and see how their efforts can be scaled. Making the switch to green steel is a good example of how to approach suppliers directly. These suppliers know their supply chain all the way down to the raw materials extraction operations, and as major contributors to the global climate crisis, they have started to work with their customers to tackle this hugely challenging problem. Developing a fossil fuel-free alternative to steelmaking is a hugely expensive undertaking, which is even more of a reason why widespread collaboration within the ecosystem is important.

### Sustainability in Action

Industry needs to find a way to join forces and motivate suppliers to invest a large amount of money in zero emissions steel that will spread the premium over multiple players and make the investment much more attractive for everyone. In the end, collaboration like this might be less expensive than offsetting or paying carbon penalties in the future.

**Dr María Mendiluce, Chief Executive Officer, We Mean Business Coalition**

We Mean Business Coalition is a global non-profit coalition working with the world's most influential businesses to take action on climate change.

### Approach 3

*Use hardware and teardown activities* to get an understanding of the complexity of products and processes and showcase these to the appropriate stakeholders in the company, such as R&D, product management, manufacturing, and logistics. This exercise will help identify low runners that could be scrapped to help speed up processes and get products to customers sooner. Any type of waste reduction also means a significant contribution to sustainability goals.

**Approach 4**
Work together with, for example, major plastics suppliers and collect ideas on how to *rewrite component specifications*. This is an important step to finding more sustainable material alternatives.

**Approach 5**
*Think recycling and circular economy!* Many raw materials are very cost intensive as well as problematic when it comes to their extraction or production, resulting in soil degradation, water shortages, biodiversity loss, damage to ecosystem functions, as well human rights violations. Where such materials have been used, recycling could prevent a lot of waste, pain, and cost. Any reduce, reuse, and recycle policy will help to minimize the negative impacts of raw materials extraction and use, but this is an area that needs more attention. It could be a good topic to explore in workshops with suppliers.

**Sustainability in Action**

We'll need effective recycling technologies, a distributed ecosystem of recycling capacities, and supportive European policy. Up to 90% of a battery can be recycled, and up to 95% of the nickel, manganese and cobalt used to make batteries can be reused. But to do this we have to have the recycling technology first for handling batteries, then dismantling, and then recovering materials that can be reused. Chinese and South Korean companies are leading the way on this, but if we want a green battery supply chain in Europe and circular manufacturing, we must get better at doing this as well.

**Peter Carlsson, Co-founder and Chief Executive Officer, Northvolt**

Northvolt is striving to produce the greenest battery in the world with a minimal carbon footprint and the highest ambitions for recycling to enable the European transition to renewable energy.

## Pushing Relentlessly

Hackathons and teardown activities held in conjunction with suppliers can create a lot of engagement and inspiration resulting in a positive, can-do attitude among all concerned. But we also know only too well, especially in the current situation where there are material scarcities and price increases, that people get dragged away into task forces to fix pressing issues that are impacting current operations. This is the default setting in most companies because it is what people have been programmed to do over the years. It is therefore crucial to stay focused on the big goal, the future pathway to net zero, and not get too easily distracted by short-term thinking. Procrastination will have potentially devastating consequences.

Time is of the essence. With corporate ambitions in mind, and with procurement teams that want to take the lead in establishing their respective roadmaps to net zero, it becomes easier to judge the effort that will be needed over a certain period of time. This is a time for pragmatism, not theory—only real steps count. Create a timeline to track progress and as a visual aid to see if more needs to be done. Focus on those categories or components that will make a real impact and push relentlessly for action and progress in those areas. Include these discussions in RFQ and sourcing processes so that decisions can be based on true total costs, including ways of mitigating $CO_2$ reductions in the supply chain, and so on.

**Sustainability in Action**

At the moment, decarbonization is a very big issue, which is why we've defined quantitative targets that include targets for scope 1 and 2, as well as scope 3, including scope 3 upstream. But how do we achieve them? We collaborate with our suppliers so we're transparent and then see what can be done to reduce CO2.

We've established a database with an external partner. From this we can derive which CO2 emissions come from which material group and from which country. That was the first step to achieving CO2 transparency. We buy steel from different countries and how does it fit together, what are the energy mixes there and how do we get that right? Additionally, two or three years ago we wrote to the 8,000 or 9,000 of the largest CO2 emitting suppliers and had discussions with them about the extent to which they are already dealing with the issue and what measures they are taking.

**Markus Strangmüller, Corporate Development & Sustainability Manager, Siemens**

Siemens is a technology company focused on industry, infrastructure, mobility, and healthcare, creating technologies and transforming the industries that form the backbone of economies.

In some cases, it will be necessary to split responsibilities between people dealing with current operations and those who are granted the freedom to work on sustainability activities, including reducing $CO_2$ footprints and improving workers' rights, activities that will ultimately create high, long-term value.

## Celebrating and Repeating

There will undoubtedly be pockets in many organizations where people have already started addressing sustainability issues and even recorded some successes. These inspirational eco-leaders must be used to encourage other people to jump on board the sustainability band wagon. Share and celebrate their stories and successes. Leading by inspiring people to be part of a positive movement is more effective than criticizing people for not doing the right thing. Of course, there will always be hardliners who will stubbornly resist change, and this is where those who support sustainability initiatives must be true to their beliefs and accept that the way forward might not be taken side-by-side. Under these circumstances, it becomes even more important to remain positive, celebrate each and every success story, and keep up the momentum.

**Summary: Set Category Work and Suppliers in Motion**

Once there has been a category assessment within the supply base, it will be clear where action is required first. Then make use one of the many sustainability tools available to cover the basics of sustainability and apply individual solutions to those categories with the highest risks. Remember, there is a wide ecosystem that can be used for support. Internally, make sure the necessary stakeholders are part of the sustainability mission. And finally, suppliers, the most powerful alliance there is, will now be aware and ready to get going on the sustainability journey.

## References

BMW. (2021). *BMW launches the Re:BMW Circular Lab*. The BMW Group. Accessed Jan 11, 2022, from press.bmwgroup.com/global/article/detail/T0338864EN/bmw-group-launches-the-re:bmw-circular-lab

Continental. (n.d.). *Continental, the circular economy*. Accessed Jan 12, 2022, from continental.com/en/press/studies-publications/technology-dossiers/sustainability/circular-economy

Ellen MacArthur Foundation. (2021). *What is a circular economy?* Accessed Jan 4, 2022, from ellenmacarthurfoundation.org

European Parliament. (2021). *Circular economy: definition, importance, and benefits. The European Parliament*. Accessed Jan 6, 2022, from europarl.europa.eu/news/en/headlines/economy/20151201STO05603/circular-economy-definition-importance-and-benefits.

Geneva Environment Network. (2021). *Earth Overshoot Day 2021*. Accessed Jan 4, 2022, from genevaenvironmentnetwork.org/resources/updates/earth-overshoot-day-2021/

H&M (2021). *H&M Group launches circular design tool "Circulator."* Accessed Jan 11, 2022, from hmgroup.com/news/hm-group-launches-circular-design-tool-circulator/

IKEA (n.d.). *A circular economy starts with a BILLY bookcase... or INGO table, or NORRARYD chair*. Accessed Jan 12, 2022, from ikea.com/gb/en/this-is-ikea/sustainable-everyday/buy-back-pubcc071810

Leahy, S. (2017). *Each U.S. Family Trashes 400 iPhones' Worth of E-Waste a Year. National Geographic*. Accessed Jan 6, 2022, from nationalgeographic.com/science/article/e-waste-monitor-report-glut

Potting, J., Hekkert, M., Worrell, E., & Hanemaaijer, A. (2017). *Circular economy: Measuring innovation in the product Chian*. PBL Netherland Environment Assessment Agency, The Hague.

Schmid, W., & Ritzrau, D. (2018). *Why the circular economy must link up the whole supply chain*. The World Economic Forum. Accessed Jan 12, 2022, from weforum.org/agenda/2018/09/why-the-circular-economy-needs-to-link-up-the-whole-supply-chain/.

Schneider Electric. (2019). *Schneider Electric wins global award for contribution to circular economy*. Schneider Electric press release. Accessed Jan 12, 2022, from se.com/ww/en/about-us/newsroom/news/press-releases/schneider-electric%E2%80%99s-wins-global-award-for-contribution-to-the-circular-economy-6024fadd4e9cb81346456af7

Sutherland, A. (2021). *Mærsk Line: Cradle-to-cradle passport for ships*. The Knowledge Hub. Accessed Jan 12, 2022, from knowledge-hub.circle-lab.com/circleeconomy/article/5910?n=Mærsk-Line-Cradle-to-cradle-passport-for-ships

UNEP. (2019). *We're gobbling up the Earth's resources at an unsustainable rate*. United Nations Environment Programme. Accessed January 4, 2022, from unep.org/news-and-stories/story/were-gobbling-earths-resources-unsustainable-rate

# Create a Movement and Drive Transformation

# 6

**Written by our Guest Author, Michaela Bürger (Supported by Nicole Ludwig)**

Michaela Bürger is the owner of Michaela Bürger Consulting and an expert in vision, strategy, structure, and leadership issues in a digital world. She is also a speaker and author. When she published her first book "Champions League for Managers" in 2014, Joe Kaeser commented "I consider Michaela Bürger to be one of the most internationally experienced and competent experts on the use of potential and team diversity." She openly and courageously tackles current leadership topics with a clear eye for the essence. With her team, she supports both corporate groups and medium-sized companies.



S. Aichbauer et al., *Responsible Procurement*, Management for Professionals,
https://doi.org/10.1007/978-3-030-98640-7_6

## 6.1 The Future of Leadership: Context, Challenges, and Necessary Skills

### The Challenge of Transformation

The transformation to a sustainable way of doing business is the running theme of this book but making this kind of change is far easier said than done. To be able to initiate change, there has to be the desire to do so and then it is a process that has to start from the top of an organization. The leadership and management of a transformation always take place in the light of collaboration among people. The way executives use their skills, talents, and personalities has significant influence on the development of the company's value and on the outcomes of the transformation.

In addition, it is vital to acquire the necessary skills and knowledge for the change. Essentials are an open environment and decision-makers who embrace change. Once the desire is there, moving successfully from A to B means having an action-guiding vision, setting strategic goals, and having the courage to try new things.

This brings to mind an experiment in Arizona in the 1990s. Four women and men, together with 3800 animal and plant species, were locked in a huge glass building, Biosphere 2, designed to be a miniature version of our planet. Owned by the University of Arizona, the building was constructed for scientists to study how Earth's living systems work (Fisher Smith, 2010).

The experiment aimed to find out whether it was possible to recreate Earth-like conditions on other planets that could ultimately be inhabited by humans. It turned out that the experiment was not successful for various reasons. It was interesting, for example, that the trees simply fell over after reaching a certain height. The reason for this was difficult for the scientists to detect. After a while, they found out that the problem was the lack of wind. The dynamics and power of the wind ensure that trees become strong and resilient. Only when the wind shakes from all sides the tree can grow roots deep in the ground. If this does not happen, the tree has no stability and falls over.

Without the forces of change acting on them many people surrender to their natural inclination to find a place in the slipstream. This saves energy. Why torture yourself unnecessarily? The reality is that there cannot be transformation without change.

Paradoxically, it is precisely uncomfortable, new experiences that allow us to achieve the goals of the future. Everything we can do, all our strengthened abilities, we owe to borderline situations. It is the winds of change and the problems that blow our way that really challenge us, make us grow, and give us new perspectives and abilities. Neuro-scientifically, no brain really wants to change. Every change forces us to leave a behavioral highway and to follow a new path that is not very well paved, unknown, and much more laborious than the old, proven one.

So, we spend a lot of energy trying to remain in a tried and tested comfort zone, hoping that the storms of life will steer clear of us, even if we constantly talk about the importance of "fresh wind" and changes that a transition to a sustainable business model will bring. However, without the winds of change all is calm and at a standstill, a state that some may welcome, but, on the other hand, much in us remains undiscovered and a transformation takes place only in words, not deeds.

Everyone has the choice of either building protective walls when the winds of change blow, or harnessing the energy they bring to transform and move from A to B.

In the business world right now, sustainability is the wind of change and moving in response to this dynamic force will make companies stronger in the long run. The roots that give companies this strength comes from their customers, investors, and employees and any company that fails to respond to this wind of change is risking the loss of these roots and perhaps collapse.

## Aspects that Call for Leadership Change

Why should people take on the hardships that every change brings? Why not just carry on with business as usual? The short answer as far as sustainability is concerned is that the growing amount of legislation is forcing companies to have at least a minimum level of compliance, but as has already been stressed throughout this book, a sustainability transformation involves so much more than ticking boxes, it is about much more, it is about the "real thing."

Almost all companies have given themselves mission statements in an attempt (that so often fails) to develop their own philosophy and to put it into words. These statements are usually developed in workshops with the executives and the supervisory bodies, accompanied by consultants.

However, there is great value to be gained from exchanging values together. Being enthusiastic or not, can inspire or not and that is what makes the difference. And employees quickly feel this. Values concerning the environment and human welfare are inherent in the definition of sustainability, and that is why leadership must factor them into the transformation process.

So, one of the main reasons to take on these hardships now is because they will lead to many companies rediscovering or creating a genuine purpose, one where they will be part of a better future. But this is going to need a new kind of inspiring, progressive leadership.

The attitude to life of an entire generation is currently changing dramatically and these people want so much more than a salary from the companies they work for—they want meaning and purpose; they want to create a better society and to live in an intact environment. They are looking for quality of life in a job with a purpose: I know why I am doing something; I understand the meaning that I can identify myself with; I feel connected in the long term; I know and live my personal values; the combination of "purpose" and values I strive for are the anchors of success. These are the types of sentiments that more and more people are looking for.

## Sustainability in Action

> I don't deny that building a new industry is not a nine-to-five job. It's tough, but we are working transparently when it comes to challenges and wins alike. We also do a lot to build a strong culture. Leadership, organizational culture, what defines us, these all play huge roles, and you have to spend time on it.
>
> From day one we took the decision to make every employee a shareholder. It's a way of letting people know that they're genuinely appreciated. That way, their competence is valued as much as capital, which seems to be a general trend. There's a lot of capital if you know where to look, but competence is much scarcer.
>
> People want to be part of creating something new and creating impact. For younger engineers specifically, the paycheck alone was not sufficient motivation but seeing the impact they could have made them join.

**Peter Carlsson, Co-founder and Chief Executive Officer, Northvolt**

> Northvolt is striving to produce the greenest battery in the world with a minimal carbon footprint and the highest ambitions for recycling to enable the European transition to renewable energy.

This has far-reaching consequences for the economy. If a company wants to survive in the future, its decision-making matrix will have to expand. It will have to formulate answers to the question of the meaning of the company and, derived from it, the meaning of the individual workplace.

## The Economic Value of Leadership and What Drives us

The value of executive behavior in companies correlates with the development of corporate value and is therefore of significant importance. The requirements for the skills of leaders in a digital world can be defined and measured.

There is a difference between talent and what someone knows and is capable of doing in their specific subject matter versus personality, or the way someone goes about doing something and manages business and people throughout the daily operations—otherwise known as their behavior. The two combined, knowledge AND behavior, reflect the value of leadership performance and ultimately in company value.

## Personal Value System as a Compass for Leadership Behavior

Our behavior is determined by our values, attitudes, and principles. They are not visible, but they influence our perception and judgments. They can be used to understand personal goals, attitudes, and principles (Fig. 6.1).

In the outmoded linear economy, one often encounters executives who describe the effectiveness of behavior, and thus its relevance, in the evaluation of executives

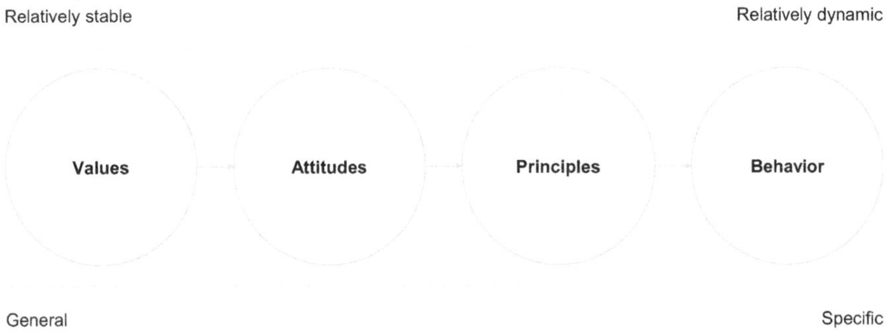


**Fig. 6.1** Mechanism of action of values on behavior

as unnecessary, even esoteric, and consequently attach little value to it. This attitude is usually expressed with the following blanket statement: "It ultimately depends on the achievement of the goals and not how this is done." This is factually right, but wrong for the long run.

Prof. Dr. Christina Hoon from Bielefeld University investigated the connection between the misuse of leadership in everyday management and the achievement of company results. The conclusion of the investigation suggested that it does matter whether there is a bad leader in a company and that poor leadership results in the overall leadership climate becoming toxic. The study also found that poor leadership filters down into other levels of management and does, in fact, cost companies money (Bormann et al., 2020).

## What Skills Do Leaders Need to Successfully Drive Change in a Volatile World?

In addition to a good constitution for everyday management, leaders need efficiency to achieve goals quickly and sustainably as a team. Equally indispensable is an interaction dynamic that encourages and spurs on others. In addition, a high level of self-responsibility is needed to manage oneself. More than ever, it is the behavior of leaders that provides orientation for their companies, their inner compass, combined with a personal authority that radiates far beyond professional authority.

All these competencies are expressed, among other things, in the behavior of leaders, which is based on their values, attitudes, and principles. *Values* are the "general guide" that steer our thoughts and actions. They are the real drivers of change. They are deeply rooted, strongly socialized beliefs of what is right and important. They are not visible, but unconsciously influence our perception and judgments. They can be used to understand personal goals, attitudes, and principles. Values are relatively stable and rarely change.

Our values and attitudes are part of our personal identity. They control our actions and our assessment competence. They give direction to leaders and help them to make good and well-founded decisions.

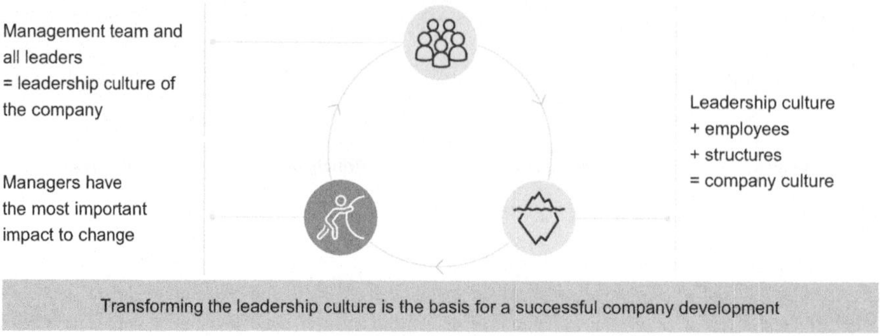


**Fig. 6.2** The most efficient lever for corporate development

The creative power of people should be put to good use by focusing on the essentials:

- deep human relationships give security and at the same time courage;
- professional and personal competence secure our skills;
- independent thinking and empowerment drive the guiding vision of change.

By focusing on these essentials, companies can grow healthily, consistently, and sustainably. It is time to think more about leadership in transformation projects, otherwise cold-hearted power games and excessive self-importance will divide us and create dissatisfaction—even though we humans could do much better.

Leadership behavior is the most effective and efficient lever of a company's value development. The sum of the behavior of all leaders can be described as the leadership culture. People behavior and existing structures and rituals and networks build the corporate culture. If the corporate culture is to be changed, for example, to react to changed market conditions or new technological opportunities, managers must change their behavior to engage people and adjust structures to ultimately secure further development of the company's value (Fig. 6.2).

Consequently, if leaders have the most important impact on corporate development, a proper talent acquisition and development process needs to be put in place to acknowledge, define, and secure the key skills of current and potential leaders.

## 7 Steps to Meet your Goals

The important question now is how can I as a manager, CEO, company owner, or supervisory board member promote value-enhancing leadership in the company and initiate a suitable change process in the organization?

After years of experience as a leader and the insights gained from many consulting projects with satisfied customers, I created a 7-step-approach. This has proven to be a successful and lasting method to change the leadership culture and one where each individual manager benefits.

1. Value creation and leadership understanding.

Develop the 10 to 12 most important values with each of the relevant decision-makers (owners, C-level managers, supervisory board members), those that characterize their personal identity. Then each of the decision-makers selects the four to six values that are considered indispensable for their personal leadership style, specifically those that characterize their leadership behavior.

Each value, which is mentioned several times by all decision-makers involved, is discussed. Three to four behavioral descriptions are created to show how this value is lived by the decision-makers. This process is repeated for the most important four to six values, the ones which the leaders consider to be indispensable in their decision-making. In the end an understanding of leadership arises according to which the decision-makers select, evaluate, scrutinize, and praise.

If there is already a leadership model in the organization, the question is how it was originally developed, and importantly, whether the leaders truly live by this model. If it was not created via the collaborative process described above, it is usually not authentic. Values may well have been written down and liked just because they sounded good or because they were state of the art in current leadership theory. The HR department then interpreted them in their rather isolated way and defined them as the company's core values. The result is guiding principles on glossy paper with little credibility and limited application in everyday life.

It is also highly recommended to explain the "why" and discuss with all the leaders the vision of the company from an inclusive and action-oriented perspective. In this way, you build on the foundation of every decision-maker and can credibly set changes in motion—both individually and collectively.

2. Involvement of management team.

It is important to inspire leaders to engage in the more detailed objectives, the planned procedures, and to use the following steps in the change program. In transformation processes, early involvement, open and regular information, awareness through guided reflection, exchange, and discussion really matter.

**Communication and Transformation Checklist**

- Objective of the project: why do we do this?
- Context explanations: values, leadership understanding, lived leadership behavior, influence on the corporate culture.
- Voices and input in the project: what is expected?
- Presentation of the procedure: how, what, and when do we implement?

(continued)

- Awareness and reflection of one's own understanding of leadership through targeted questions and group discussions.
- Exchange within the group of participants.
- Presentation of the values that are indispensable for leadership on the part of top management.
- Discussion of the leadership model and the desired leadership behavior.
- How do we live up to and measure the lived behavior of leaders?

3. Individual analysis of leadership behavior (target/actual).

Only if the managers can and want to implement change will the desired change process in the company succeed. From our point of view, the core element of the approach is therefore a biographical interview with two external consultants, which reflects the behavior of the manager with regard to the requirements for sustainable leadership in the future.

A biographical interview ensures the most reliable conclusions about personality traits through structure and questioning technique and is also the most effective form of questioning in which people are the focus and not just the object of analysis.

In addition, the interview promotes the beginning of a reflection of one's own work and leadership behavior.

4. Survey about leadership culture.

Get an idea of the leadership culture in your company. Use an online culture monitor to ask your employees (and thus also your managers across all levels) how they perceive the existing cultures of cooperation, creativity, ambition, and control in the company and which culture they would like to see in order to achieve the best performance in a motivated and successful manner.

Here, the understanding of leadership described in point 1 is decisive for the desired direction of the development of the leadership culture.

5. Individual and collective leadership development.

As a result of the individual analysis of leadership behavior and the survey about leadership culture, in addition to the personal assessment of each individual manager, you will also receive a collective assessment of leadership behavior and can specifically develop your executives and leadership behavior in the company both individually and collectively.

6. Individual clarification of values.

Our values determine our behavior. As we grow and develop, we are shaped by our environment and learn what is right and what is wrong. Each of us, and thus every leader, has unconsciously developed values that we strive for, and these are manifested in attitudes, principles, and daily behavior. They provide guidance and thus serve as a compass for life.

To enable change, it is therefore important that your leaders know which values are behind their behaviors so that they are able to reflect on and deal with themselves, with other people and with challenging situations more easily, effectively, and with less conflict. This is the basis for the realignment in a change process, which is necessary for all managers.

Only if we know where we came from can we set off bravely in new directions, and head into the unknown, without becoming unfaithful to ourselves.

7. Building and integrating leadership development.

Anchor the understanding of leadership in the company through the annual evaluation of professional skills and leadership behavior in the context of the employee appraisal process. This allows you to identify progress, standstill, or regression. It is also interesting to note to what extent goals were achieved. What does it mean if there has been above-average achievement of goals and moderate leadership behavior? The effects on employees can very often only be observed years later when entire teams collapse. Then management will be caught by surprise because everyone is under the impression that achieving business goals must go along with great leadership and well-being.

But that is not the case. Pay careful and clear attention to the correlations and then put your teams together. It is often better to get rid of managers who do not have the right leadership skills or put them into positions where they can be their best expert in their field.

This is the only way to build up consistent succession planning in the company that is sustainable and offers perspective and choice.

This is the only way to keep very good managers in the company and have the guarantee that everyone has the ambition to become better and further develop every day.

This is the only way to attract talent that supports driving change and wants to learn and develop.

This is the only way to always have the right leaders, at the right time, in the right place and thus steadily increase the value of the company.

From human value to company value. That is the way. Have fun! (Fig. 6.3).

So finally, we would like to encourage you to follow this path, because for us it is quite clear that leadership value, employee satisfaction, and company value have a clear correlation. A good example on how to become successful by following this approach is the Berger Fahrzeugtechnik GmbH in Austria.

> Under the BERGERecotrail® brand, we manufacture semi-trailers and chassis structures in lightweight steel construction. The minimized dead weight increases the payload and brings the ecological and economic requirements in road transport down to a common denominator.

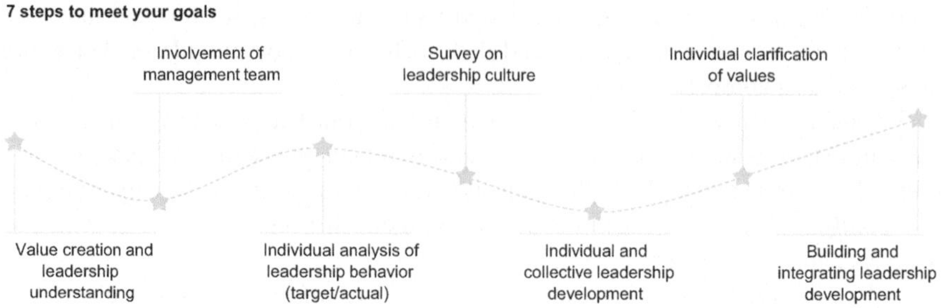


**Fig. 6.3** Seven steps to meet your goals

Due to its continuous growth process and the ramp up of the new production facility in 2019 the company was facing a needed change. In order to conduct this, a leadership team with a versatile and diversified skillset was required. On the one hand the team needed to share the vision and strategy of the company and on the other hand the defined cultural values. Therefore, the leadership team had been empowered in a way to take care of a high goal-efficiency, good interaction skills, a clear attitude by knowing their values and principles and a distinctive self-responsibility. The analysis by a biographical 6D-Leader process helped to ensure that the team members were able to reflect the current behavior compared to the necessary capabilities to be successful when making decisions and driving innovation.

We believe that an innovative organization needs to have leaders at every level, not just at the top, in order to stay competitive. However, the leadership team, as well as the company, is subject to a learning curve and constant development process. We have enormous growth potential, which we can only realize if we have the best leaders who keep our purpose and the achievement of goals in mind in their daily actions and who steer and develop employees and customers in such a way that we become better every day. Leadership behavior is just as important as professional competence and experience. We have made a tremendous push and are on a clear, strategically well-thought-out path into the future.

As a team, we have achieved a significant increase in performance over the last two years. Not only do we feel comfortable (or more comfortable than before), but we are also performing better.

**Bastian Litterscheid, Managing Director, Berger Fahrzeugtechnik**

**End of guest author's contribution**

## 6.2 Engage and Empower Leaders

### How to Become an Inspiring Leader and a Role Model

Leadership behavior has a significant influence on the achievement of goals and the development of the company's purpose. Leaving aside the obvious ability to steer their companies to profitability, leaders need a constitution that carries them through their day-to-day management, but on top of that they need to be able to give clear, unambiguous guidance to the people they work with, especially in a complex and volatile environment. In order to achieve goals quickly, good leaders should be able to motivate the people around them and spur them on towards their individual and collective targets.

There are certain traits that good leaders have in common, characteristics that will make the difference and conjure something magical in the companies and people they lead. Good leaders:

- empower teams and give them the mandate and responsibility to take decisions,
- bring clarity into the business and prioritize tasks when teams are in doubt,
- create a learning organization in which teams feel safe to dare and learn from failure,
- give constructive feedback to people for them to understand, reflect, and be able to change,
- are empathetic to the needs, feelings, and thoughts of others and are able to orchestrate diverse people from different backgrounds and working units,
- persist in taking care of people and do not let daily operational issues drag them away from it,
- develop and nurture new competencies in the teams, such as embracing change, building relationships, orchestrating supply chains, and challenging the status quo,
- recruit the right people with a can-do-mentality.

## Even the Best Leaders cannot Do it Alone

While leaders play a key role, no transformation of any kind will be successful if there are only leaders, no matter how inspiring and visionary they are. All successful transformations start from the top, but the entire workforce has to clearly understand why it is necessary and what they have to do. In a sustainability transformation, there has to be a change of mindset on top of the requirements for sustainable products or services, support for a more circular economy, more focus on a sustainable supply chain, emissions reductions, as well as equality, inclusivity, and diverse teams internally and among suppliers. That is a lot to change, but all of these will have a positive effect on attracting new talent and like-minded people who actively want to work for companies that have a strong background of sustainability. And in turn, those employees who do not agree with the new vision will sooner or later leave the company.

The magic happens when a forward-looking, modern leader meets and works with people in an organization who are open to sustainability. That is when leaders can truly make a difference by winning the hearts and minds of a willing workforce so that they can start to tackle sustainability-related issues together.

## 6.3 Excursion: Inner Development Goals Offer a New Approach to Old Problems

> How is it that, despite all the experience, power, technology, money, and knowledge that humankind has accumulated, we still seem unable to defeat the biggest challenges facing us today? Climate change, pollution, mass migration, overpopulation, corruption, disease and pandemics, extremism, slavery, war, terrorism, drug trafficking, hunger, weapons proliferation, species and habitat loss, prejudice and racism, unemployment, water scarcity, antibiotic resistance, human rights abuses, poverty, illiteracy, infant mortality, and inequality.
>
> **Simon Anholt, Chair of the Steering Group, Inner Development Goals Initiative**

### Are we Able to Defeat the Biggest Challenges Facing us Today?

The Sustainable Development Goals that are referenced throughout this book were created in 2015 as a comprehensive plan for a sustainable world by 2030. The 17 goals cover a wide range of issues that involve people with different needs, values, and convictions. There is a vision of what needs to happen, but progress has so far been disappointing. An important part of the reason for this is that we lack the inner capacity to deal with our increasingly complex environment and challenges. In essence, we cannot solve the problems we face with the same mindset that created them in the first place. It was this thinking, backed by extensive research, that led to the creation of the Inner Development Goals (IDG), an initiative that provides a blueprint of the capabilities, qualities, and skills we need in order to achieve the 17 SDGs.

In these interview extracts, Simon Anholt, Chair of the Steering Group, Inner Development Goals Initiative, and Jan Artem Henriksson, the initiative's Chief Executive Officer, share their thoughts and ideas on how we can effect a change of mindset by developing the inner abilities we now all need to tackle the issues we face. The IDGs are a way to educate, inspire, and empower people to be a positive force for change in society and as such they are directly relevant to company leaders who are seeking to sustainably transform their businesses.

### Collaboration Can Build a Better World

> The problem is the fundamental culture of governance: the idea that leaders are only responsible for their own people and their own slice of territory. Today it's clear that they are responsible for every man, woman, child and animal on the planet; for every inch of the earth's surface, the atmosphere above it, and the ground beneath their feet. And if they don't like the sound of that, they shouldn't be leading anything. In short, all of us—not just leaders—need minds that telescope, not minds that microscope.
>
> Every challenge facing humanity today is caused by human behavior, so if we want to change the world, we have to change ourselves. Yet we still educate people in ways that only made sense before humanity became interdependent, and its problems interconnected. We still encourage and reward behavior that produces local benefits and extraneous harm. We

still put competition ahead of collaboration and cooperation, even though the three mix beautifully. We all want to help, but we have so much to learn before our help can be truly productive before we can exercise *real agency*.

**Simon Anholt**

## We Need an Updated Human Operating System

What skills do we need to develop to implement the UN's 17 Sustainable Development Goals more effectively and more quickly? And how can this be achieved at the individual as well as the organizational level? This question is at the core of the Inner Development Goals Initiative.

I don't believe that the same people that we are today will be able to reach the 17 United Nations SDGs. We need an update on our human operating system to get there. We will not be the same people when we truly collaborate on these goals. There is a blind spot in the SDGs because we're not looking on the individual and collective inside of our personal skills and the culture that we are creating. That's why we truly need to use the most effective interventions for human development to help us grow, so we can reach the SDGs within the time frame that we have.

The Inner Development Goals (IDGs) speak about a big 'us' perspective, it's all of us, not just the politicians or the leaders. We all need to embrace a developmental perspective. But the leaders we choose often do not have the best skills and capabilities to collaborate with other people on complex issues. Sometimes we choose people who are very confident or explain things in very black and white terms. That can give us some comfort and a sense of false security. But that will not get us closer to the SDGs. So, we need to choose those leaders or politicians who have more of the qualities that we need globally to succeed.

From the research that I have seen, only a minority of the leaders do that today. It is a huge challenge, but we need wiser, more compassionate, holistic leaders. The good news here is that these skills can be developed. It's something that we can grow throughout our entire lives with a lifelong learning and growth.

**Jan Artem Henriksson**

## An Urgent Shift of Perspective from the Private Sector

We're convinced that building skills and competencies is an essential part of making this crucial shift happen. The overwhelming response to our initiative suggests that we're on the right track: although the IDGs were only founded just over a year ago, Costa Rica has already become the first country to commit publicly and officially to organizing its public sector around the framework we identified in the first phase of the initiative. We are now talking to several other countries interested in following Costa Rica's example.

But this isn't only about countries: we believe that a shift of perspective is urgently needed in the private sector too. In particular, we see opportunities in education and business to help achieve the Sustainable Development Goals by focusing on inner development. From board members to employees, inner growth must become a top priority for companies. It's an approach that helps leaders to appreciate a wider range of human perspectives, to understand and benefit from cultural differences, and to perceive people's true needs in order to build resilient partnerships for cooperation and collaboration. We all need to get fit for the race ahead.

**Simon Anholt**

### Structural Changes Are Long Overdue

> Our aim, via the IDGs, is to create ecosystems that break down barriers between sectors and between people from different backgrounds. We believe that the time for important cultural and structural changes in companies is long overdue, and that they need to start asking the questions, 'What is the true purpose of companies in an age of global crisis? How can we rethink financial flows in a way that accelerates instead of obstructing our progress towards the Sustainable Development Goals? Where can we recruit the talents, we need to meet the challenges ahead, and how can these talents be best developed?'
>
> **Jan Artem Henriksson**

### Relate to each Other, and the World

> We're at a moment in history when the future of humanity is in the balance. Our only hope is to focus more on the values and behaviors that bring us together, and less on those that are tearing us apart.
>
> **Simon Anholt**

## 6.4 Leadership and Learning to Make a Difference

There are many shining examples of purpose-driven companies that have understood how they can play a role in actively tackling some of the most pressing issues of the world by pushing for the cultural and structural changes that are necessary. In a sense the timing could not be better in that all players in the ecosystem are pushing in the right direction: consumers, by thinking more carefully about what they buy and who they buy from, are forcing the business world to act more responsibly; widespread and growing concerns about the environment and human rights abuses have resulted in mounting regulations around $CO_2$ and workers' rights along supply chains; entire countries and companies have signed up to net-zero scenarios; investors are challenging businesses to focus on long-term and sustainable ambitions when selecting where to allocate their vast financial resources; and young professionals expect their employers to have a clear purpose that ingrains sustainability in everything they do.

These are powerful and compelling forces and good leaders will need to quickly learn how they can be harnessed and put to good use in all areas of business. The "right" leaders can be strong influencers and play prominent roles in the transformation process. However, as well as being technically competent in their fields, good leaders need to learn continuously and be open minded if they are to instigate the transformations that will take their companies into a future focused on sustainability.

The progressive leadership that has been discussed here has influence beyond its immediate surroundings. The beauty of it is that it stretches beyond the boundaries of the company into all kinds of business scenarios, be it with current and potential suppliers, other partners, or the entire ecosystem beyond the four walls of the company. This type of progressive, open-minded, and forward-thinking leadership is precisely what is needed to bring a sense of clarity and urgency to the most pressing issues facing the world. It is the type of leadership that will undoubtedly be found heading each of the more than 2000 businesses and financial institutions that are working with the Science Based Targets initiative (SBTi) to reduce their emissions in line with climate science (SBT, 2021). The leadership of the many partners and collaborators that make up the We Mean Business Coalition (WMBC, 2022) is also characterized by an outlook that extends beyond their own immediate surroundings, as are the top managers from the hundreds of companies that are making use of the best-in-class climate action initiatives offered by WMBC.

There are only positive outcomes to this type of leadership—companies headed by these people attract and maintain the right talent; they futureproof themselves against sustainability-related legislation; they are more favorable options for investors; and finally, mindful customers prefer to do business with them. All gain, no pain.

## References

Bormann, K. C., Hoon, C., Graffius, M., & Hansen, C. (2020). A trickle-down model of abusive supervision and firm performance in family and non-family firms. *Academy of Management Proceedings (Vol. 2020, No. 1).*

Fisher Smith, J. (2010). Life under the bubble. *Discover Magazine*. https://www.discovermagazine.com/environment/life-under-the-bubble

SBT. (2021). *Companies taking action, science based targets*. Accessed Nov 22, 2021, from sciencebasedtargets.org/companies-taking-action

WMBC. (2022). *Leading companies are going all in and seizing the opportunity of climate action. We mean business coalition*. Accessed Jan 4, 2022, from wemeanbusinesscoalition.org/business.

# Outlook 2030 and Conclusion

# 7

## Where Do you See the Topic of Sustainability in 2030?

I see that this is a movement that will continue to intensify over the next decade. The question is if it will move fast enough to reach the set targets for combatting climate change and other parts of the SDGs. To be successful, we simply can't afford major disruptions. [...] If we look only at the rapid technological developments, the overall circumstances at companies, and their willingness to transform, we have a good foundation for achieving what is needed by 2030.

**Fredrik Gustafsson, Senior Manager of Sustainability Transformation, Cordial**

Cordial is a management consulting firm based in Stockholm, Sweden. Its main focus is strategy, transformation, and business development with sustainability as one of the key areas.

My hope is that it will be something that we talk less about because it'll already be such a natural part of how we work and do business. This would mean that we've reached the tipping point where we're finally starting to go in the right direction globally.

**Henrik Larsen, Chief Procurement Officer, A. P. Møller-Mærsk**

A. P. Møller-Mærsk is an integrated container logistics company and member of the A.P. Møller Group. With a dedicated team of over 80,000 people, they enable global trade for a growing world.

I believe that climate change will remain highly important over the next decade. By solving the challenge of climate change we'll also cover many of the other challenges related to environment, such as air pollution, over consumption, deforestation, and biodiversity. Human rights and especially human rights issues in the supply chain will also keep gaining traction.

**Chaojun Li, Global Head of Sustainability, WS Audiology**

WS Audiology develops, manufactures, sells, and distributes hearing aids. WS Audiology's ambition as a global leader is to unlock human potential by making wonderful sound part of everyone's life.



S. Aichbauer et al., *Responsible Procurement*, Management for Professionals,
https://doi.org/10.1007/978-3-030-98640-7_7

The biggest issue for the next couple of decades will be climate change and the required decarbonization. It's fundamental and already galvanized. Policymakers and NGOs will demand that companies in any sector of any size prove that they have reduced carbon emissions by credibly reporting on their demonstrable actions. If corporations fail to do so, they will come under immense reputational and regulatory pressures. So, it is in the corporations' interests to move fast and in earnest on this issue. Linked to this, as well, is the other big development in that we're going to be increasingly moving away from linear, purely extractive industries and extractive industries-based value chains to increasingly circular economy-based value chains. This is an overdue development.

**Dr. Nicholas Garrett, Co-founder and Chief Executive Officer, RCS Global Group**
RCS is a proven leader in data-driven ESG performance, responsible sourcing, and responsible mining assurance.

Renewables are more competitive than coal so we hope market forces will win out, but it needs time to make that transition happen. Coal has no place in the world. Oil's place in the world is also diminishing as the uptake of electric vehicles accelerates. There'll still be some used in aviation and shipping but eventually there'll be alternative fuels in these sectors as well.

Although gas produces half of the emissions of oil it still produces emissions. Because of the current high energy prices people think they need to invest more in gas production, but we should actually be investing more in gas storage because the same storage facilities could be used for hydrogen in the future. Natural gas pipelines could also be repurposed to help transport hydrogen.

The future is renewables and green hydrogen, and we need an infrastructure that supports them.

**Dr. María Mendiluce, Chief Executive Officer, We Mean Business Coalition**
We Mean Business Coalition is a global non-profit coalition working with the world's most influential businesses to take action on climate change.

As we said in Sect. 1.1, it is time to change the way we treat the planet and all its ecosystems. It is time to change the way some people are abused for the sake of lowering the cost of consumer goods. It is time to change the way we live, time to change the way we do business, and time to put sustainability on top of the agenda when making supply chain and procurement decisions. The world in 2030 should be a cleaner, more equitable place than it is now. This will be the case if the highly informed insights from the pioneering sustainability leaders featured above and throughout this book come true.

The quotes above are a clear indication that attitudes about sustainability, and the increasingly prominent role it plays in procurement, will have changed dramatically by 2030. In line with the UN SDGs, the aim is to end poverty and hunger, equally value the human rights of all, achieve gender equality, and ensure the lasting protection of the planet and its natural resources. If equal importance is attached to the economic, social, and environmental aspects of sustainable development, they could go a long way in helping us to solve many of the problems that human activity has caused. Thankfully, an increasing number of people, governments, and companies agree with this and are acting accordingly.

One point came across very clearly throughout our research for this book. Whether we were talking to seasoned professionals from a variety of industries who have been championing sustainability in their companies for some time, dedicated people from the non-profit sector, entrepreneurs from start-ups, or energetic, determined young procurement professionals, all agreed that boosting sustainability-related actions needs to happen now.

However, it is the younger generation—the future workforce, managers, and leaders of the corporate world—that is proving to be one of the strongest and most determined drivers of sustainable development. In the course of developing the ideas and gathering the details for this book, we took every opportunity to speak to the new generation of young procurement professionals. Not surprisingly, they are ambitious and strive for attractive roles, decent salaries, and good career development opportunities. But they also want their jobs to be purpose driven, they want to know that what they do makes a positive difference to the people and natural world around them. For them, sustainability has been an important element of their jobs since their careers began.

To attract the right talent from among these young professionals, the procurement function will have to change so that sustainability is an inherent part of the role. Sustainability has to be clearly visible in procurement's strategy and targets. This "movement" needs the support of top management as well as being anchored across the organization and other functions. Young professionals are also digital natives, so they understand how transformation works and how digital tools can be applied to all aspects of their lives, including tackling the business challenges of sustainability.

Procrastination must be replaced by proactivity. This is especially true where decarbonization is concerned and the goal of staying within the 1.5 °C temperature rise target, which leading scientists say is essential for our survival and the survival of future generations. We are in a race against time and the clock that will determine our future is ticking loudly. It is a sound that can no longer be ignored. Procurement has a prominent role to play in helping us to win this race. It can be the function that really takes control of the corporate sustainability agenda by re-inventing itself so that sustainability criteria are on a par with cost efficiencies and quality when it comes to purchasing decisions or the choice of which suppliers to partner with.

This is the essence of this book. Define a purpose for procurement that incorporates sustainability. Build a challenging strategy and roadmap and systematically anchor sustainability in all relevant processes throughout the organization. Create transparency about the environmental, social, and ethical dimensions of the supply chain by monitoring all relationships with suppliers and partners across the value chain. Digital tools are invaluable here just as they are for helping to understand the supply chain beyond tier 1—it is generally the smaller, lower-tier suppliers that carry the most risk. And when developing measures, focus on high-risk issues first and work *with* suppliers to create joint roadmaps. Collaboration and co-creation, internally with cross-functional teams, and externally with partners and suppliers, are two of the keys that will unlock many sustainability challenges. Finally, a sustainability transformation is a marathon, not a sprint, so find a rhythm, maintain energy and dedication, and keep going. The rewards will be worth the effort.

Above all, please act *now*.

# Sustainability in Action: Interviews with Experts

# 8

To round off our guide to achieving sustainability in procurement we have compiled some of the interviews that were conducted during the writing of this book, which demonstrate what sustainability in procurement looks like in the real world, in a variety of industries.

From battery developers to audiology to shipping giants and more, the interviews present a selection of different approaches to sustainability, company-wide and for procurement specifically. The common thread throughout is the notion of the importance of sustainability in procurement. Having now read our guide, use these interviews as inspiration to take action, embed sustainability in your procurement function, and turn ambitions into impact.

Peter Carlsson, Co-founder and Chief Executive Officer of Northvolt
Nils Eichberger, Vice President of Procurement, LBBW (Landesbank Baden-Württemberg)
Sakuto Goda, Principal, Nomura Research Institute
Nina Göntgen-Voss, Director Sustainability, Forto
Fredrik Gustafsson, Senior Manager of Sustainability Transformation, Cordial
Thomas Janvier, Vice President of Procurement, INNIO
Simon Juul Toft, Lead Specialist—Green Procurement, Ørsted
Henrik Larsen, Chief Procurement Officer, A. P. Møller-Mærsk
Chaojun Li, Global Head of Sustainability, WS Audiology
Dr María Mendiluce, Chief Executive Officer, We Mean Business Coalition
Markus Strangmueller, Corporate Development & Sustainability Manager, Siemens
Thomas Udesen, Chief Procurement Officer, Bayer and Co-Founder, Sustainable Procurement Pledge



S. Aichbauer et al., *Responsible Procurement*, Management for Professionals,
https://doi.org/10.1007/978-3-030-98640-7_8

## 8.1 Peter Carlsson, Co-Founder and Chief Executive Officer, Northvolt

The future supply chain is going to require more partnerships as well as transparency.

### About Northvolt

Northvolt AB is a Swedish battery developer and manufacturer, specializing in lithium-ion technology for electric vehicles. Its mission is to build the greenest battery in the world with a minimal carbon footprint and the highest ambitions for recycling to enable the European transition to renewable energy.

### About Peter Carlsson

Peter Carlsson is the CEO and founder of Northvolt. He left the position as CPO and Head of Supply Chain at Tesla Motors in the end of 2015 to become an entrepreneur, advisor, and angel investor. In 2016, Peter founded Northvolt with the mission to build Europe's largest lithium-ion battery factory.

Before joining Tesla, Peter was the CPO for NXP (former Philips Semiconductors) responsible for purchasing and outsourcing. Located in Singapore, he was a country manager for Singapore, Malaysia, and Philippines. Before starting at NXP, Peter was CPO at Sony Ericsson since 2005. Peter had a total of 13 years' experience from different roles with Ericsson and Sony Ericsson. He has a master's degree with specialization in production and quality control at Luleå Technical University in Sweden.

### Sustainability in Action

**You had a high-impact job at Tesla, orchestrating its whole supply chain. Why did you start Northvolt? What was the trigger point?**
I left Tesla at end of 2015 but did not really plan to start a company. I wanted a break and started to invest in start-ups, both in Silicon Valley and Europe, and I joined a number of boards.

Then I realized two things. First, it is much more fun to build something on your own, instead of just giving advice. After talking with a few people and looking at Europe I was then contacted by Business Sweden with the question "What would it take to get Tesla to build a battery factory in Sweden?" Then I looked at the material and realized that some education is required in Europe to get the facts straight for a battery business in Sweden.

The second thing I realized is that investors were open to this idea. We agreed to do a concept study about building a battery factory in Europe.

There were two other factors that made this idea appealing. First, there's tremendous pressure to reduce carbon emissions, especially in the power generation and auto industries. We put together a small group of four people in Silicon Valley. Intrigued by the idea, we started to travel and meet stakeholders in Asia and Europe to see if it was feasible to start a battery factory in Europe. The more time we spent on this, the more we recognized that there is a big need.

We realized that the supply chain for batteries is very fragmented with players that do not fully disclose how energy-intensive battery manufacturing is. When you look at the supply chain, seeing it from the perspective of mining and raw materials processing, you quickly realize that making cells is incredibly energy intensive. If you took a vertical integration approach, you could change the game. For example, if you responsibly source a battery's "active material" (the chemically active components of the two electrodes of a cell and the electrolyte between them, including nickel, manganese, cobalt, and aluminum), and put it into an energy grid based on renewable and low-cost energy, then you could compete with low-cost labor countries.

**What finally convinced you to start a battery business?**
After 6 months of this pre-investigation, this was a bit too interesting not to try. A little bit too important not to try. Even though battery manufacturing is very capital-intensive . . . you need to go big or should not even try.

**Did you find investors who shared your enthusiasm for battery manufacturing in Europe?**
Yes, we went out to investors, raised $15m, recruited 25 people, and put the shovel in the ground within 18 months. Since then, assumptions have changed, but the fundamentals have stayed the same. We underestimated how many people we needed, how much funding . . . but that is what you do when you start-up a company.

**Were you worried about the Asian battery giants?**
In the beginning, we thought the best way to start this was to form a tech partnership with one of the big players. We spent a lot of time thinking this through but eventually recognized we could spend years, and secondly, recognized how different the cultures are. At some point we skipped the idea of having one big partner in favor of having many and building a strong team ourselves. This has worked out. Now we have a number of key partners in the equipment, construction, energy, and transportation sectors. There are also a number of companies that are willing to co-locate where we build.

The EU had been clear that as part of building up the European industry in step one, they want to build sustainable batteries and have a green battery label. The EU wants to create a framework for green certification and customs tariffs and taxes on the $CO_2$ of the product. This is all under discussion. The timeframe for a potential labeling is around 2023.

The fact that the EU is driving this has pushed competition and raw material suppliers to look more actively at the EU and invest here.

**What you have done is impressive and you could easily be described as a role model for other companies wanting to make a green transformation. What is the secret behind getting Northvolt started?**
We managed to build a strong case around the pilot plant. We went out to 10 companies, all potential customers, and offered them the chance to participate or invest in some type of partnership. Almost all were excited about the idea and the opportunity to build up this competence in Europe. Then we entered the corporate world of governance and processes and of course, what we thought would be a much faster process, turned out to be substantially slower.

**Would you agree that within companies there needs to be more alignment between procurement, R&D, manufacturing and possibly even the M&A departments, that companies should adopt more start-up-like behavior?**
Getting started took much longer than we thought. At first, we had more individual projects, but then had our first customer with ABB, then Vestas, then Scania and from these got the proceeds we needed to invest the 250 million euros on the pilot plant in Västerås.

Everyone was very positive about the new angle in the industry. Most people were unaware how energy intensive this industry would be. If you build up a plant with coal-based energy, the problem is only pushed to another place.

There were two other things that might not have been on our radar when we first started but turned out to be highly beneficial:

First, when we decided to do the vertical integration of active materials, suddenly, instead of having a fragmented supply chain, Northvolt could combine the sourcing of those materials with its direct requirements to specific mines and get clear transparency down to the mining level.

Second, building up a whole new industry in a sustainable way attracted a lot of talent and specialists from around the globe. People want to be part of creating something new and creating impact. For younger engineers specifically, the pay-check alone was not sufficient motivation but seeing the impact they could have made them join.

**You have highlighted two important ingredients for building a sustainable company: transparency and collaboration in the wider ecosystem. What is your experience of the general commitment to sustainability in European countries?**
There are very few companies that are transparent about their carbon footprint. They produce nice reports, but does everyone fully understand what impact they have?

Some KPIs are important: cost per kwh, quality, capacity … but so is sustainability. When we first started to talk about this, no one listened. Four years later, it has finally become important for companies and also for German customers. In my experience, it is best not to start with compromises—if you cannot build a factory under similar $CO_2$ conditions as in Sweden, do not build a second factory in a country with a high $CO_2$ energy mix.

**Is it fair to say that Northvolt is a pioneer in that it is nudging the political arena for Europe to be part of a growing industry and building sustainable prosperity?**
When we started to look at Europe, we talked with industry professionals, and knowing that they demand just-in-time concepts, there would have to be regionalization. We knew there was a case for building up capacities right here. We made the European Commission aware of the fact that Europe would have its own ecosystem, and that this would have a huge impact on R&D, equipment, component supply development, universities, and that the impact on the economy would be so much bigger if they were to support this. The European Battery Alliance was created out of this in 2017 and now has 500 member companies.

**Are you happy with the support of the Commission?**
In the beginning the European Investment Bank (the bank of the European Union) provided financing, which was great. More recently this has become less of an issue because investors are reallocating their portfolios towards the green transition.

Now the focus is on the supply chain and how to build the competencies in R&D, manufacturing, and so on in a short period of time. There are some important topics that we are working on together with the EC and its European members.

To produce truly green batteries, we need to manufacture them using renewable resources, such as hydro power, combined with a low carbon footprint around.

We are looking to find cobalt from non-Congo sources, and this could mean new ways of working with the raw materials and minerals.

Again, to reduce the overall carbon footprint, we should build strong supply chains within Europe. Developing European machine builders in equipment making is also important for strategic competitiveness.

And let us not forget about recycling and the circular economy. These are vital parts of green manufacturing.

All these topics are under discussion.

**How do you embed circular thinking into the organization, the design team, R&D, and throughout the entire company?**
Recycling is an important part of green manufacturing and it directly benefits production too, because it means that the raw materials we need for new cells can be replaced with recycled materials.

We will use more and more batteries in Europe in the coming years so we have to build an infrastructure to recycle them. We will need effective recycling technologies, a distributed ecosystem of recycling capacities, and supportive European policy. Up to 90% of a battery can be recycled, and up to 95% of the nickel, manganese, and cobalt used to make batteries can be reused. But to do this we have to have the recycling technology first for handling batteries, then dismantling, and then recovering materials that can be reused. Chinese and South Korean companies are leading the way on this, but if we want a green battery supply chain in Europe and circular manufacturing, we have to get better at doing this as well.

**Northvolt is now in the tough phase of implementation, industrialization of the product, and preparing the plants for day-to-day manufacturing. How do you motivate your team during this busy period?**
I do not deny that building a new industry is not a nine-to-five job. It is tough, but we are working transparently when it comes to challenges and wins alike. It is very important to celebrate the wins.

We also do a lot to build a strong culture. We have got people from all over the world working here, and that means different leadership and different approaches to resolving problems. Having a homogenous workforce might mean getting things done quicker, but for long-term success try to build a diverse culture in all its aspects. Leadership, organizational culture, what defines us, these all play a huge role, and you have to spend time on it.

From day one we took the decision to make every employee a shareholder. It is a way of letting people know that they are genuinely appreciated. That way, their competence is valued as much as capital, which seems to be a general trend. There is a lot of capital if you know where to look, but competence is much scarcer.

**What is your final advice to the procurement organizations where sustainability is concerned?**
The future supply chain is going to require more partnerships as well as transparency. Data sharing is going to be a key enabler for long-term competitiveness. Hardware and software will both be important. People will have to develop the ability to learn from data to design better in the first place and enhance products out in the field. This will involve the entire supply chain and procurement must orchestrate this transformation.

## 8.2 Nils Eichberger, Vice President of Procurement, LBBW (Landesbank Baden-Württemberg)

> Find fun in the subject. It's very meaningful [. . .] the most important thing is to understand sustainability within the overall context of the business.

### About LBBW

LBBW is a Mittelstand-minded universal bank with deep roots in Baden-Wuerttemberg. We are a partner for medium-sized companies, listed groups, institutional customers, savings banks, and retail customers.

## About Nils Eichberger

Nils is an experienced Vice President of Procurement with a demonstrated history of working in the banking industry. His areas of expertise lie in negotiation, IT strategy, strategic sourcing, management, and risk management. He is also passionate about digitalization, climate protection, sustainability, and employee experience. As an EVP at LBBW he is currently responsible for Group Procurement and Security.

## Sustainability in Action

**What is your personal connection to the topic of sustainability?**
I have two children and I am convinced that we all have a very personal responsibility to live in a resource-optimized or climate-positive way, since we only have one Earth. We have to be economical with what we have available to us. For many people there is also an opportunity to drive the issue forward in the companies we work for.

**When did LBBW decide to really make an effort on sustainability? What triggered this decision?**
This is actually intrinsically in LBBW's DNA. We first addressed the issue in 1991, 30 years ago, when we published what was then called the first environmental report in Germany. We are a company under public law, with the city and state as our sponsors, and I would like to say that we have the claim that we act sustainably as a company. For us, sustainability also means our responsibility as a bank and trust center.

For more than 4 years now it has been one of our four strategic directions and is defined through Scopes 1–3. Sustainable development is important to our company, which is why we have sustainable human resources development and look to recruit sustainability-minded people. Getting the right people means we will be successful in the long term.

**When you talk about sustainability in LBBW, what scope do you consider?**
At the moment ecological sustainability is very much in the public spotlight, at least in the daily press and beyond, so regulatory issues related to climate risks are now a definite concern for us just as they sold be for all financial service providers. But we also take our social responsibilities seriously as well as the 17 SDGs

**As LBBW, what is your leading framework? You mentioned the 17 targets.**
Yes, we have everything, also a sustainability strategy. We have signed the Principles for Sustainable Investment, which are based on the UN's 17 goals. We have also started and joined several different initiatives.

**Who oversees sustainability at LBBW? How would you describe the organizational setup?**

We have a head of sustainability who reports directly to our CEO, who has overall responsibility for sustainability. In the end, it is a team sport, and everyone is responsible to some extent. We use different approaches to sustainability in our product portfolio. For example, we have sustainability investment products. To a certain extent we can control where funds flow through a process of supply and demand such as in companies that are subject to certain standards and thus promote sustainability.

We also have financing that is linked to sustainability KPIs of targets of the companies. In other words, the more sustainable the company's development, the more favorable their credit rating.

We also pay close attention to our own operational emissions. Everyone who works at LBBW is made aware that they can have a positive effect on the climate by doing or not doing certain things and they are incentivized to do so.

We had a sustainability fair more than 2 years ago. We managed to share and discuss about sustainability with 280 colleagues for a day: with fair booths, with external partners, but also, very importantly, with workshops, where colleagues could participate, where they could contribute their own ideas for the further development of the subject along these four lines of action, i.e. planet, credit, carbon footprint, and personnel. Naturally, this generated a lot of momentum, and we did a bit of advertising for the topics and acquired a lot from this, and then at some point it becomes a bit of a self-runner.

But we also use sustainability as an opportunity to expand our "business model." For example, we have developed a $CO_2$ dashboard for LBBW based on the demand and consumption information from purchasing, which lowers our $CO_2$ footprint throughout the organization. Every employee, every organizational unit has the opportunity to see and control their emissions. In the expansion, we now distribute the $CO_2$ targets for operational emissions in a balance score card.

**How do you deal with sustainability when it comes to procurement?**

This begins relatively globally at the supplier or partner level, where we have been working with a sustainability agreement for 10–15 years that simply includes certain criteria. We have a two-stage onboarding process and potential suppliers have to satisfy a minimum number of criteria. For example, there is the sustainability agreement that partners have to commit to, which is also audited and, in case of doubt, leads to exclusion if it is violated. This is the basic level, and then we have category-specific requirements, which is where we switch from supplier to category management. Individual products then go for certification to Blue Angel, the ecolabel of the Federal Government of Germany. With consulting or professional services, we expect people to travel by train instead of by car or by plane. This is similar to what we ourselves expect of our employees and where we define requirements for products or services that we purchase.

**Do you offer your suppliers any form of bonus if they are particularly sustainable?**

No, not yet. But operating sustainably is simply mandatory if you want to work with us. However, we recently held discussions with our category managers about whether it makes sense to include a sustainability day, where we award suppliers who support our sustainability goals, or who, from our point of view, perform extraordinarily well. The idea is to promote what they are doing as an example to other companies in our community. We have had to sideline this idea for the moment because we are very busy with other activities, but to answer the question, there is no bonus.

**Compliance with LBBW's sustainability requirements is audited. Do you use a tool to do this?**

We can conduct audits but we are currently reorganizing ourselves in this respect because the large number of regulatory and compliance agreements are actually stressing out the entire organization.

The supply chain law is waiting for us. Of course, we are looking at this from a number of different angles. On the one hand, it directly affects us, and on the other hand, we are actually looking at it again under the heading of Beyond Banking to see to if it could become part of our business model in relation to the issue of trust. It is all about the inheritance of sustainability criteria along the supply chains. We are currently looking at whether it would be interesting to work with a customer in this area.

**Do you have any tool in place specifically for sustainability?**

Yes, we have. Our sustainability report, which deals with our carbon footprint, among other things. We have a solution that collects all the consumption data for the entire organization.

**What tools do you use to get transparency into your supply chain and to ensure that LBBW's standards and requirements are met?**

That is where we have just changed. We are doing this through our supplier management. In the meantime, we have joined Prospeum, a small start-up that I personally think has come up with a very clever idea and we are currently migrating our previous solution to it.

**What are currently your biggest challenges?**

What we keep looking at, and what we are concerned about in some areas, is World Savings Day. It is still the same as it used to be. It is still very classic. The kids come with their piggy banks and there are gifts. The whole topics of promotional gifts, marketing items, textiles are already topics that give me such a bit of a stomachache. We define requirements and certificates from the suppliers. Nevertheless, I look at it with concern, because at the end of the day, of course, I do not know whether what I am holding in my hand is sustainable or not. I have it on paper, but you never know.

**What recommendations would you offer to other (procurement) organizations that are starting on the sustainability journey? What have been key success factors at LBBW?**
Find fun in the subject. It is very meaningful. At least that is how I feel about it. Sustainability is a topic that procurement would sometimes want to push away and instead refer it to other departments such as compliance and outsourcing management. But sustainability should be seen more as an opportunity to develop beyond procurement into new business models. When someone is just starting out, I think the most important thing is to understand sustainability within the overall context of the business. This will show where the opportunities lie. At the end of the day, no company can ignore sustainability nowadays, so embrace it.

There are cost implications, of course, but ultimately it has a direct bearing on reputation and how the company is viewed in the market. There have been enough studies to show that future generations of employees will be very concerned about the sustainable impacts of the companies they work for and less concerned with fringe benefits such as company cars. They want meaning and purpose, and I think that is important here.

**In an ideal situation, with the resources, money, and courage that you need, how do you see the LBBW's sustainability journey to 2030?**
To me it is about developing things forward. New business models, how procurement behaves in the networking of suppliers with each other and their further development. I would like to see a strong community, where we drive sustainability forward together. There is already a lot of that in the form of platform exchanges, and pitches. That is the direction I am looking in. The other things are also important, you need that too. But we can still develop the world together with our suppliers and our customers.

## 8.3 Sakuto Goda, Principal, Nomura Research Institute

> I hope that nobody is talking about sustainability then because it has become a matter of course, because it has become normal to everyone.

### About NRI

Nomura Research Institute (NRI) is a leading consulting firm and systems integrator from Japan with expertise in the formulation of strategies, globalization of business operations and planning.

## About Sakuto Goda

Sakuto Goda is an experienced principal consultant at the Nomura Research Institute based in Tokyo. With over 10 years of expertise in the areas of strategy and development he uses his experience and comprehension of industry trends to help customers develop strategy, support decision-making, and form and execute effective business measures.

## Sustainability in action

**What is your perspective on sustainability in Japan? What is the current status of the topic?**
Sustainability has become prominent in Japan. Generally, one could say that the current movement related to sustainability came originally from Europe, but I can clearly see that it has become more and more important in Japan. Both in private and in business people are constantly talking about it. I often hear my family talking about sustainability and especially about waste reduction or waste management or $CO_2$ reduction. It is also a political topic and is therefore quite prominent on TV. It is interesting to see how fast the topic has emerged.

**What is sustainability for you?**
Currently, carbon neutrality is by far the biggest topic when it comes to sustainability—at least that is my personal view. The reason for that is quite simple. Prime Minister Suga, as well as the environment minister, announced quite ambitious reduction targets after they were appointed to office. In fact, the newly announced goal is to reduce emissions by 46 percent compared to 2013. Not surprisingly, carbon neutrality is currently the hottest topic in Japan.

Besides that, I would say that waste reduction is also one of the biggest topics. It gained importance through news about the problem of plastic waste in the oceans. Those images went viral very quickly. This has led to several changes in private life, like the fact that people have to pay for plastic bags in supermarkets.

Although $CO_2$ emissions reductions and waste management are the most prominent topics, I would define sustainability as being much broader than that. For me sustainability also comprises social aspects, water treatment, and labor rights, for example. However, in Japan, it seems that most people view sustainability as mostly concerned with the environment.

**Do you see sustainability as a trend or something that will bring changes to Japanese society?**
I would like to describe my opinion by using an example. A few months ago, I talked to my five-year-old son who is currently going to kindergarten, and he was talking about SDGs. He told me that he learned about SDGs in the kindergarten and wanted me to tell him about my personal opinion on this topic! The younger generation is

starting to focus on sustainability quite early. That is why I do not think that it is just a trend but something that will always be around us.

I also think that Japanese culture has always comprised at least some aspects of sustainability. We always had discussions about the concept of sustainability, although we might have called it something different, but in the end, it meant the same. In Japanese we use the word "Mottainai," which basically describes the regrets we have for something that was lost or the waste produced by using something. I think that the word sustainability is now the chance to incorporate the meaning of the existing expressions into our daily lives—just in a different way.

But it is not only about the way people think. Sustainability has also changed the traditional way companies do business—or at least is about to do so. Japan is quite famous for its manufacturing industries and it is that sector that needs to further reduce waste or to rethink the traditional way resources are used. If they do not, these companies will be going out of business due to increased regulatory requirements or customers' demands for sustainable alternatives in the future.

**Do you see differences when looking at Japan compared to other Asian countries? If so, which ones are the front runners in sustainability?**

It is difficult for me to give a detailed answer to that question. In fact, I think that all countries are facing similar issues when it comes to sustainability and that all countries need to act. I do not think that there is any country with a significant lead over others. It feels that we are all at the same point with equal problems and it is more a case of learning from each other.

Of course, I hope that Japan will become a frontrunner in terms of sustainability in the future as today's concept of sustainability is already embedded into the Japanese way of living. That is why we need to make sure to leverage this asset and to develop new business models around it.

**Do you see sustainability as a pressing topic among your customers?**

Yes, over the past years, the number of sustainability-related enquiries has significantly increased—from multiple industries. Most of these enquiries ask for support with how to respond to the increasing demand for sustainability and how to react. Besides that, many companies want to leverage sustainability to develop new business models. As an example, companies that really want to optimize carbon emissions within the supply chain need transparency, optimized services, and new equipment. In other words, companies do not only want to react but to capitalize on new business opportunities.

**Has it been more of a top-down engagement (coming from management) or a bottom-up (demanded from the employees) that has triggered the increased focus?**

I would say that this depends on the type of company and the customers this company has. However, in most cases, I see the topic coming from the top and then going down. Sustainability is a relatively new topic in Japan and there is no concrete route for actions. Managers need to develop new action plans, define new

KPIs, and come up with new business opportunities. This cannot be promoted bottom-up. For that you need management that defines the way forward.

**What are the main challenges they are experiencing and focus areas they are interested in?**

As I said earlier, I think that uncertainty is one of the key challenges. There is no clear pathway leading to the right actions and there is a lack of solutions for all relevant sustainability problems. It is about finding solutions, developing new business opportunities, and leveraging what is out there to find your individual way through the sustainability jungle.

Besides that, I think that the economic factors of sustainable business are another key challenge. Today, sustainability is associated with relatively high costs, meaning that becoming sustainable, or developing new sustainable business models, requires huge investments. Most managers are looking for suitable ways to monetize new business models while keeping additional costs to the minimum. Some people think that customers are willing to pay more if products or services are sustainable. I think that this is only partially true and depends on the nature of the product. From what I see, people might be willing to pay a small surplus for sustainability, but in most cases this surplus does not fully mitigate the additional costs of offering the product or service. So, in the end, it is a business case for each company and each product that managers need to decide on.

A third challenge is resource limitations. The sustainability concept is global. People talk about it on a global scale and try to develop concepts that have quite a broad scope. However, the implementation of these concepts requires local or regional action, meaning that each action has to be executed within one country, region, or factory. The challenge associated with that is the limitation of resources. To give you an example, in Japan there is insufficient land for solar energy or wind power, but people still ask for renewable energy sources. So, the question is, how do we do that? It is about a shift in resources, but if you are not rich in the required resources, it is very difficult to become truly sustainable.

**Do you see digital tools as an enabler for positive change? If so, how are Japanese companies are using the tools?**

Definitely. Digital technology is for sure one of the key enablers for sustainability. However, technology alone is not able to answer sustainability challenges, as most sustainability topics deal with physical challenges. Think about the concept of the circular economy where products and materials are re-entering the production process through recycling and upcycling with the aim of keeping them within this circle for as long as possible. So, in the end, digital technology is not the full answer to sustainability challenges. It certainly paves the way as it helps to generate transparency, define your ecological footprint, or to track resources. So, I would say that digital technology should be a prerequisite as an enabler of sustainability but is not the silver bullet to solve all the challenges.

**What do you perceive as prerequisites (skills) to ensure that sustainability is successfully embedded within a company's operations?**
Generally, I think the first required skill is a change of mindset. But this is mostly true for the older generations. Coming back to the example of my five-year-old son, I see the younger generation treating sustainability as a matter of course. It is the older generations where this mindset shift needs to take place. So, a five-year-old kid already knows about SDGs and wants to hear my personal opinion on it. This shows that it is simply a matter of time before that mindset shift will be visible.

I also perceive sustainability as a transformation, meaning that people require transformational skills such as being open minded and constantly looking for innovation opportunities. In the business context this means that managers need to educate and motivate employees, clearly define the need for action, and compensate peoples' efforts to become sustainable in an economical way. In the end, it goes hand-in-hand with the mindset shift—managers need to ensure that this mindset shift also takes place within their companies.

**What advice do you give to your clients who do not know where to start when it comes to sustainability?**
First, I would advise them to define where to start and to assess the scope. It is about understanding where a company can contribute, which products to focus on, where the biggest footprint is, which technologies can be leveraged, and what business opportunities to focus on. So, it is about generating transparency among existing processes and the goals that the company aims for.

Secondly, it is about mindset. Managers need to ensure that everyone is going in the same and correct direction. Goals and ambitions must be communicated openly, and the vision needs to be clear and understood by everyone.

Thirdly, it is about co-creation. Companies need to speak to their suppliers and outside stakeholders to define new ideas and bring them to life. It is about working together to see where sustainability can be leveraged and how new ideas can be developed together. One company on its own will most probably not be able to change society. However, by joining forces and through learning from each other, great ideas can be developed.

**Where do you see sustainability in 2030?**
I hope that nobody is talking about sustainability then because it has become a matter of course, because it has become normal to everyone. This vision means that processes have been adopted and that new business models have been developed while existing models have been adopted according to the requirements of sustainability-related transformation.

However, I think that, in reality, we will see targets that were defined for 2030 will need to be adjusted over time. I am afraid that we will see companies and also countries, in let us say 2027, adjusting their targets. In the end, I think that it highly depends on the younger generation. They are more attuned to sustainability and will also be more aware of what it means. In the end, it will be about finding new technologies and leveraging the trend to do better than we are doing today. I really hope that my vision for 2030 becomes reality and I put a lot of hope in the next generation.

## 8.4 Nina Göntgen-Voss, Director of Sustainability, Forto

> We see ourselves as change agents for the logistics industry. We've already turned the industry upside down once by providing more transparency through digital solutions—even though people did not clearly express demand for what we're offering. The same thing could and will happen with sustainability.

### About Forto

delivers a highly transparent, frictionless, and sustainable digital supply chain. Their platform addresses the entire process stream, supporting customers with greater visibility, insight, and control.

### About Nina Göntgen-Voss

As a Director Sustainability, Nina Göntgen-Voss is responsible for Forto's strategic direction with regard to sustainability, which is an integral part of the company's vision and underpinned by numerous initiatives for and with employees, customers, and partners. After graduating from the EBS University of Business and Law as well as internships in start-ups and consulting, Nina joined Forto in May 2016 as one of the very first employees. Since then, she contributed substantially to building the company, working in various departments and leading several cross-functional projects and teams.

### Sustainability in action

**To what extent is sustainability part of your company agenda?**
Sustainability is at the core of our company purpose: we are in business to create a more responsible society by fostering both sustainability and prosperity, which are closely interlinked.

We fundamentally believe that global trade is something good, especially if you consider how much wealth and prosperity stems from it. Different nations can participate in the global economy which enables all nations to increase their productivity—given that there is equal access to global trade. In Asia alone, more than a billion people have been lifted out of poverty due to global trade in the past decade. After the introduction of containers and "containerization" over 50 years ago, the digital transformation of global trade is now the next major step. Therefore, we believe technology is a key driver for global prosperity.

At the same time, we also see that global logistics and global supply chains have a major impact on emissions. Cargo transports account for around 6–7% of global greenhouse gas emissions. It is our joint responsibility to make a difference and curb climate change. We must transform global supply chains in a sustainable way to secure the livelihood and prosperity of future generations. Therefore, sustainability is fundamentally anchored in our company's purpose and part of our daily business.

**When did Forto decide to really make an effort on sustainability? What triggered this decision?**
We started the business in 2016, full of energy and with a thousand topics to deal with at once, sustainability being one of them. Enabling a sustainable logistics industry is tough when you consider the different standards and regulations and its global scale. So instead of doing many things just partially right, we set our focus on digitalization in the beginning, and are now also putting full attention on the topic of sustainability.

As mentioned, cargo transportation accounts for 6–7% of global greenhouse gas emissions. Forto is part of this problem, but we can also be part of the solution and create significant positive impact. Most of all, we want to take important steps in the right direction and inspire others—both organizations and individuals—to join us on this journey.

**When you talk about sustainability at Forto, which elements of sustainability do you consider?**
Our sustainability understanding is constantly evolving. We started with the standard framework that includes Environment, Social, and Economic sustainability. By now, we understand that it is not just three separate pillars that have some intersection, but rather that the planet we live on with its conditions such as stable temperatures, nutrition, etc., is the basis for life in all its forms in colors. And social stability in turn enables economic systems with competition and profits to exist.

Due to the global climate emergency we are facing, we developed our sustainability strategy with a strong focus on the environment in mid-2020. However, environmental, social, and economic issues cannot be solved in separation. There is a reason why, in 2015, the United Nations defined 17 interlinked Sustainable Development Goals to achieve a better future for all people and the planet—including goals like zero hunger, no poverty, quality education, life on land, climate action, responsible consumption, and many more. To achieve a viable future for all people on a healthy planet we cannot leave any problem unattended. Therefore, we are planning to address problems beyond environmental issues as well—again starting with a strong focus first until we are able to diversify and extend our efforts.

**Forto initiated the "Climate Action Group." What is that?**
Originally, our co-founder Ferry Heilemann triggered actions to build a more sustainable company and about 2 years ago his passion for sustainability led to the

formation of the Climate Action Group. It was founded to give everyone in the company the opportunity to join in and contribute towards sustainability at Forto. Since we deeply believe that doing the small things consistently can make a big difference, it started small. First ideas included reducing paper use, changing electricity providers, and rethinking office nutrition. Now, for example, we only offer organic vegetarian and vegan meals.

**What does sustainability mean for the logistics industry?**
Just a couple of months after we started our internal Climate Action Group, we participated in a panel discussion on sustainability in logistics. We shared our thoughts on how digitalization can support sustainability in logistics. The feedback from the audience was overwhelming and motivated us to bring our ideas to life.

We see ourselves as change agents for the logistics industry. We have already turned the industry upside down once by providing more transparency through digital solutions—even though people did not clearly express demand for what we are offering. The same thing could and will happen with sustainability.

**What specific sustainability-related offers do you have for your customers?**
We are seeing a very strong increase in our customers' interest in sustainability, and we appreciate them taking important steps by developing fully-fledged sustainability strategies. It brings great joy to everyone at Forto to support them on this journey and educate on why this topic is important—for all of us.

Our offer is derived from our sustainability strategy with strategic objectives related to emissions—we are working on achieving 100% climate-neutral operations both internally and service-related. A standard emissions strategy follows the sequence from generating transparency first, to then reducing emissions, and finally offsetting what cannot be reduced.

Since effective emissions reduction options are not available at scale within the logistics sector yet, the standard emissions strategy cannot be applied. As a transformational tech company in a hard-to-abate sector, we therefore chose to combine offsetting and transparency as a first step. Offsetting allows us to take immediate action with immediate impact while also starting conversations on sustainability with our customers. However, offsetting is only a stepping-stone, not a long-term solution. To achieve true change and reduce emissions we need technological and behavioral changes.

Forto's Unique Selling Point (USP) is that we give our customers control through transparency—initially, for logistics, but transparency is also a necessity on the path towards sustainability. Our offer currently includes transparency about the emissions generated by shipments, voluntary $CO_2$ compensations for all modes of transport, and carbon-neutral sea LCL shipments at no additional cost.

**What have you achieved so far?**
Initially, our offsetting product was not actively requested by our clients. We had to be the ones to start the conversation about sustainability. We listened, explained, educated, and discovered the problems and possible solutions together as a team.

We also trained our sales team on our offsetting offering and sustainability in the logistics industry. In the very beginning, we had a few cases where clients wanted to learn more about our sustainability offering but we could not communicate our messages clearly. Now, our sales staff are well informed about sustainability in logistics, everything we do as a company, what we offer at this moment in time, and what we plan to offer in the future.

Within one quarter after the official launch of our first sustainability product, 64% of our customers with committed sea freight volumes have joined in for our offsetting program, which translated into 38% of the related volume. In Q3, that number rose even higher to 50% of sea freight FCL volumes booked with offsetting. It was an astonishing success. We made a difference by being transparent and by explaining our goals and values. And we did all this while freight rates were increasing tenfold, there were shortages of capacities, and an overall increase of traffic on the oceans.

Keeping in mind that offsetting is just a first step towards sustainable logistics, we are committed to exploring greener transport modes. For example, 90% of our German-based Rail Pre- and On-Carriage intermodal volumes are transported using trains powered by renewable energy by default.

**Awareness and motivation are key drivers to launching actions that can reduce emissions. What do you recommend to other companies that want to do something similar? What have been the key success factors at Forto?**
Connect with others and meet people on an emotional level. Use pictures to transport your key message. Make use of the various trigger points. Everybody can be taken on a journey. You just have to find the right way to do it, and also listen to others and take in different perspectives, opinions, or concerns. Sustainability is a very complex topic, but we are all in the same boat—and we strongly believe in collaboration over competition to make businesses futureproof while enabling a viable future for all people and our planet.

**What is Forto working on offering customers in the future?**
We are planning to direct more effort to emissions reductions as part of our emissions strategy.

Since you cannot reduce what you cannot measure, we are currently working on finalizing our real-time emissions API in collaboration with our partner Planetly. This becomes the basis for visualizing emissions data across our platform, creating in-depth emissions reports, identifying emissions reduction potential and thereby enabling data-based sustainable decision-making.

To reduce emissions, we are in the process of launching pilot projects where biofuel is used as an alternative to fossil fuels. In the future we want to offer transport options with lower emissions by default with the eventual goal of zero emissions.

Moreover, we are currently looking into potential solutions linked to the Supply Chain Act. We keep a close eye on existing and upcoming regulations to be a knowledgeable and helpful partner to our clients. But we will not stop at what is legally required. Our goals and our motivation go way beyond that and double down on our efforts to make supply chains fully sustainable.

## 8.5 Fredrik Gustafsson, Senior Manager of Sustainability Transformation, Cordial AB

Rules and regulations that match sustainability targets are crucial. It needs to be made more difficult not to contribute than it is to contribute to a more sustainable future.

### About Cordial AB

Cordial is a management consulting firm based in Stockholm, Sweden. Its main focus is strategy, transformation, and business development with sustainability as one of the key areas.

### About Fredrik Gustafsson

Fredrik is currently a senior manager at Cordial AB where he helps his clients with sustainability transformation. He comes with broad experience in both consulting and the corporate sector where he has worked as head of sustainability for the international railway operator MTR Nordic and as a sustainability manager at the Swedish pension company Skandia. He also worked for more than a decade for Accenture as a management consultant with a focus on sustainability.

### Sustainability in action

**What is your focus at Cordial?**
Our strong point is transformation and in particular helping clients get buy-in for their strategies within their organizations. Many companies have an idea of what they want to do but are missing the pieces in their operating model to truly integrate sustainability. In addition, sustainability, especially in combination with digitalization, makes innovative, new business models possible. Our long experience of working in this space, together with our razor focus on successful transformation, make us a prominent advisor in sustainability transformation.

**Was there a trigger point or a certain time when Swedish companies started to ask more about sustainability?**
The interest in sustainability has exploded in the past 2 years in Sweden. One of the big differences we are seeing is that today every business leader is talking about sustainability. Before, it was more often managed within a smaller team within the

company, but now most companies have understood that sustainability is a strategic question that requires top management attention.

In Sweden, Greta Thunberg has certainly had an impact when it comes to bringing these questions higher up the agenda, along with other sustainability "milestones" such as the Paris Agreement from 2015. The pandemic has played a role in speeding up the process, but at the same time the ball was already in motion. We could especially see this among consumers, where we, for example, noticed a reduction in air travel in Sweden already before 2020.

**Would you say that a majority of companies are motivated by avoiding negative publicity and risks, or by the possibility of using sustainability as a competitive advantage?**

We do not necessarily see that it is one or the other, and it definitely depends on which industry we are looking at. But overall, I would say that the current rules and regulations are not the driving forces of sustainability. Larger companies have much higher demands on sustainability than what is required by regulation. In the supply chain, for example, suppliers face tougher requirements on both disclosures and actual progress. One motivating factor is the fear of being left behind. There is a general understanding that sustainability is rising in importance and if you do not act swiftly, it is only a matter of time before you start losing your market share, or regulations catch up and your company becomes obsolete.

Pressure from consumers is also helping to push sustainability in B2C industries. When consumers make up their minds on what is good or bad, companies within that industry need to adapt quickly to stay relevant.

**Has it been more of a top-down or bottom-up engagement that has triggered the increased focus on sustainability?**

We can see that it is starting to become an integrated part of companies' strategies so it is a top-down engagement. It is not possible to make a separate sustainability strategy, it needs to be an integral part of the overall business strategy.

**Do you see that there is a stronger focus on the environmental part of sustainability and that the social and governance parts get less attention?**

I do believe that the other parts of sustainability also get attention, but the climate topic is for sure top of the agenda. Scope 1 and 2 have been measured by many companies for quite some time now but if you want to make a proper assessment of your Scope 3 within the framework of the GHG protocol it becomes much more complex and requires collaboration. Most companies need their suppliers to start the same journey as them, which is a massive undertaking. Since all companies, one way or another, are part of a bigger context in a supply chain, all companies need to make this their focus. Before processes and tools have been fully implemented to measure this, we will most likely continue to see a focus on the environment.

**Do you see a risk that sustainability work could become a desktop exercise rather than something with a lasting impact, especially related to the work of keeping track of the supply chain?**

Most companies are not doing their own audits but instead rely on external parties to do this for them. But for industries that have traditionally been under scrutiny for sustainability, for example, the retail industry, their own commitment and controls are usually more detailed and go further into the tiers of the supply chain. This kind of transparency and traceability in the supply chain is something that will only gain in importance across all industries over the coming years.

**Is there an equally strong focus on making use of digital tools to help increase the transparency and traceability in companies' supply chains in Sweden as there is in other parts of Europe?**

Yes, there is definitely an increased focus on traceability and the wide range of digital tools available makes this easier. Technologies such as blockchain are used in more and more industries to improve the quality of this work. On the other hand, we also see an ongoing discussion on the attempts to reduce risks in the supply chain. That includes risks related to social aspects such as child labor and human rights issues, but also risks related to climate change.

Covid-19 was also an eye-opener for companies that had a high geographic concentration of suppliers. We are now starting to see companies diversifying their supply chain locations and we might even be on the verge of companies starting to in-source vital parts of their production. Even if this is a decision that often means an increase in cost, many companies have realized that the market is more and more ready to pay for the increased certainty and reduced risk.

One example of this is the Swedish company Synsam that specializes in eye health and eye fashion. They recently decided to change their business model by bringing home their entire production of glasses from China to Sweden and consequently changed to more of a premium provider. At the same time, they became a provider that lives up to the highest sustainability standards, providing local employment in their main country of sale and gaining full control over their supply chain.

**Do you see a need for skill enhancements within organizations to successfully go through the sustainability transformation?**

Larger companies that have been focusing on sustainability for a longer time often already have a sustainability team and ambitious targets. For them, the real change and transformation is where they need to focus to accomplish these targets, and educating the organization is part of that. In some cases, the knowledge of how to get there might not even exist today, so that is why it is important to encourage innovation, collaboration, and new partnerships as ways of reaching the targets. In Sweden, there is a country-wide initiative called "fossil-free Sweden" that has developed roadmaps for different industries so that the actors within an industry can come together and work towards a common goal. One example is the journey towards fossil-free steel, which requires a joint effort from the steel supply chain together with energy companies who need to provide the industry with green energy.

**Are there specific industries that are leading the way, or that you see will be moving fast in the coming years when it comes to sustainability?**
The industries with the biggest emissions, such as the energy sector, are the ones that are moving fastest at the moment, and this will most likely continue in the coming years since they need a rapid shift to renewables. The same goes for industries like transportation and automotive where the entire business model is shifting from selling cars to mobility, and where there is a boom in innovation with hybrid and electric vehicles.

**Where do you see sustainability in 2030?**
I see that this is a movement that will continue to intensify over the next decade. The question is if it will move fast enough to reach the set targets for combatting climate change and other parts of the SDGs. To be successful, we simply cannot afford major disruptions. Risks in the shape of political conflicts and war have the potential to slow things down or even stop the positive progress that has now started. That is why I believe the 17 SDGs are a great framework for demonstrating the complexity of sustainability but also showing how all parts need to come together to ensure a sustainable future. If we look only at the rapid technological developments, the overall circumstances at companies, and their willingness to transform, we have a good foundation for achieving what is needed by 2030.

**What would help speed things up further?**
Rules and regulations that match sustainability targets are crucial. It needs to be made more difficult not to contribute than it is to contribute to a more sustainable future. There is also a need for more international measures. Sustainability goes across borders so international agreements and laws are needed, such as global taxes on $CO_2$. If this could be achieved, the transition would follow automatically.

Something that we tend to forget in discussions about sustainability is the "reduce" concept and how it has a big impact on the bottom line. There are lots of ongoing discussions about innovation and how to find new ways of doing the same things as before. The fact is, however, that any reduction of resources that we can achieve, be it energy usage, consumption, or by applying the circular economy, goes straight to the bottom line and will have an immediate positive impact on sustainability.

## 8.6 Thomas Janvier, Vice President of Procurement, INNIO

We don't see any contradiction between achieving our sustainability targets and generating savings. [. . .] Stop reading this book and get started today!

## About INNIO

INNIO is a leading provider of renewable gas, natural gas, and hydrogen-based solutions and services for power generation and gas compression at or near the point of use. With the Jenbacher and Waukesha gas engines, INNIO helps to provide communities, industry, and the public access to sustainable, reliable, and economical power ranging from 200 kW to 10 MW. INNIO also provides lifecycle support and digital solutions to more than 53,000 delivered gas engines globally, through its service network in more than 100 countries. They deliver innovative technology driven by decarbonization, decentralization, and digitalization to help lead the way to a greener future. Headquartered in Jenbach, Austria, the business also has primary operations in Welland, Ontario, Canada, and Waukesha, Wisconsin, USA.

## About Thomas Janvier

Thomas has been the Vice President of Procurement at INNIO since 2017. An engineer by education, Thomas spent most of his career working for GE in different locations across Europe and North America. His interest in sustainability started 15 years ago when he decided to change his career to something where he could better use his skillset to make a true impact. This landed him in the renewables sector within GE, and he has since then continued to work with sustainability both within procurement, and in his roles within operations where he got to apply lean and six sigma methodologies. With its core goals of maximizing value while minimizing waste, lean has a lot of commonalities with sustainability and is something that Thomas now applies in his role as head of procurement at INNIO.

## Sustainability in Action

**INNIO regards itself as "an important contributor to the sustainable and greener energy solutions of today and tomorrow." How is sustainability anchored in the company?**
Sustainability is embedded within every part of the company, starting with our products. As an example, we are currently collaborating with HanseWerk Natur in Hamburg to commission engines which can run on 100% green hydrogen. You can also clearly see it in the way we manage our production sites. Here in Jenbach where topics like recycling are given a high priority, the energy comes from renewable sources, such as renewable electricity, and we are installing solar panels on our roofs. It is also visible in the job roles and responsibilities; we have, for example, appointed a sustainability, risk, and compliance leader for the procurement organization and we have a senior leader responsible for diversity across the INNIO Group.

**When did INNIO decide to really make an effort on sustainability? What triggered this decision?**
There was no official start date, but it gained more traction when we became a stand-alone company in 2018. However, even before this, sustainability was high on our agenda through initiatives like our ISO certifications. In 2021, we received the EcoVadis silver medal rating,[1] and this is, of course, something we want to keep and improve on.

**When you talk about sustainability in INNIO, what scope do you consider?**
It covers all parts of ESG, including the full scope of the GRI standards and the 17 Sustainable Development Goals. We also conducted a materiality assessment that helped us focus on four pillars:

- *Products*—All engines should be ready to be powered by 100% climate-neutral gases by 2025, they should not emit any climate-relevant methane by 2030, and 100% of the materials used to manufacture the engines should be recycled, renewable, or reclaimed.
- *Environment*—Here greenhouse gases are the focus since other environmental aspects are already well established. The target is to have production sites reduce their GHG emissions by 50% by 2030 and we want our suppliers and distributors to commit to the same targets.
- *Employees*—We want diversity to increase by 25% by 2025, diversity and inclusion training should have been held for at least 70% of management annually, and employee engagement should reach 85% by 2025.
- *Society*—Here we will continue to support social initiatives and actively engage with our local communities.

**What do you do to ensure a sustainability mindset throughout the organization?**
We do not want to leave it to chance and think that the right people will come to us automatically and that people will be aware of our initiatives without it being communicated. It all starts with our cross-functional Sustainability Review Board, led by the Group Head of Sustainability, which sets the vision and drives the message. We also have announcements from our INNIO President and CEO and success stories related to sustainability that are communicated regularly. So, there is a top-down effort to create awareness. We got very positive feedback from both internal and external stakeholders and employees when we joined the UN Global Compact, this sends a strong message that we are ready to act now.

In addition to these conscious efforts, INNIO is lucky that we generally attract people that are sustainability-conscious and intrinsically motivated to drive sustainability. We believe that this ties back to the industry we are in and that our

[1] Since the time of this writing, INNIO has achieved Gold rating with EcoVadis.

different facilities are located close to nature so people can work with a sustainability purpose and live their lives close to nature.

**How does the sustainability review board work?**
The Sustainability Review Board consists of leaders from different functions. I am part of this panel for procurement and the board reports directly to the board of directors at INNIO. Each cross-functional member of the board cascades its messages and actions to their respective function.

**How does your team keep on track of the latest developments?**
This also starts with the Sustainability Review Board. Their responsibility is to make the organization aware of regulations and general trends. In Procurement, we arrange monthly training workshops and send out targeted communications to different roles to ensure that everyone has the knowledge they need.

**How are sustainability practices integrated within procurement?**
Procurement plays a critical role in our sustainability work. We have created a roadmap of actions for procurement, including updating our code of conduct, CSR Guidelines, and other documents we use with our suppliers. We make sure that all parts are connected to the global agreements that we follow, and we have recently started collaborating with EcoVadis to further improve our practices.

As for other roles, there is a sustainability sourcing leader in the procurement team, but it cannot be a one-person-show. We are breaking down the actions to the different parts of procurement and ensuring that each person in the team knows what sustainability means to them and how it is integrated within their work. This integration, and the integration within our existing processes, is something that we will implement this year. When we talk about supplier performance, the supplier quality engineers will be responsible for auditing the suppliers with regard to sustainability data—this comes naturally to them since they already have similar responsibilities for quality.

**What roles do your strategic buyers play in your sustainability agenda?**
Our strategic buyers are the ones integrating sustainability clauses in our supplier contracts. We have some standard language regarding ESG in our general terms of purchase, but we believe in agreeing on individual targets based on what the material aspects of sustainability are for each specific supplier. The strategic buyers also work together with the sustainability manager to develop specific targets for their categories. In Indirect, for example, we identified energy efficiency levels for real estate and leased properties. For direct categories, we will partner with Engineering to reach 100% sustainable material inputs. We are currently setting the baseline for this and we will then agree on measures to take.

**How do you engage with your suppliers to match your targets?**
We have always been involved in responsible procurement topics such as human rights, conflict minerals, and REACH. In recent years, we have broadened the scope

to include, for example, recycling and $CO_2$ emissions. The engagement takes different forms, but one is that our supplier scorecards now have a sustainability section, so our suppliers must have a clear sustainability agenda in order to get full scores.

**Do you go to tier1 of the supply chain or further when it comes to ensuring a sustainability agenda?**

For high-risk categories such as conflict minerals we go all the way to the source, in this case the mine. That being said, our gateway for this information are the tier1 suppliers, since it is with them that we have a contractual relationship. They are the ones that need to take the commitments further down the tiers of the supply chain. We are also a member of Responsible Minerals Initiative.

**What happens if a supplier does not fulfill their targets?**

The supplier is less likely to be awarded new business. This is something we are just starting and are working on creating awareness, but in the future, we plan to set a minimum requirement on the score that they need to achieve to get new business, and possibly also to continue doing business with INNIO.

**What contribution does EcoVadis make along your supply chain?**

EcoVadis is our partner for rating our suppliers. We started with our largest and most strategic suppliers and will expand from there. We are also partnering with EcoVadis to measure our supply chain's carbon maturity and we will use them for ESG risk mapping along the supply chain in the future. We already have a silver rating and have received improvement opportunities that we are aiming to add to our action plan, so we will be able to move to a gold, or even platinum, rating in the future. This new action plan will be included in our sustainability report that will be published shortly[2].

Individual suppliers must pay a fee to EcoVadis to get rated, but they can use the rating for all sorts of purposes, including for other customers (not just INNIO). We have had pushbacks on this from some suppliers, but acceptance is increasing as more and more companies become aware of the benefits of this standardized rating.

**What tools do you have in place to ensure your standards and requirements are being met along the supply chain?**

On Conflict Minerals and REACH, we use a dedicated service provider that supports with tracking, gathering responses, and arranging corrective actions.

**Can savings and sustainability targets go hand-in-hand?**

We do not see any contradiction between achieving our sustainability targets and generating savings. This is mainly because sustainability in many cases means saving resources, which in the end also means saving money. Connecting back to

[2]INNIO's sustainability report has been published in Q4 2021.

the concept of lean—the reduction of waste is something that also goes hand-in-hand for both sustainability and savings.

**What have been the key success factors for INNIO?**
The personal commitment of key members in the company has played a key role. That is why it is important to communicate your plans clearly to get the buy-in you will need.

**How do you see the topic of sustainability developing in the next 10 years?**
I expect that we will continue to adjust and work with our current goals that stretch to 2030. Globally, we see that the world is already behind on where we need to be, so we need to continuously check if we are moving fast enough, and if not, move our efforts up a gear.

**What would help you to accelerate your sustainability efforts?**
Active dialog with our stakeholders is one of the ways, be it regulators, customers, shareholders, employees, or communities. We are there to deliver for our stakeholders so if they voice a need, we will listen. It is the same for procurement, which is the customer of our suppliers, and is therefore responsible for voicing the need for sustainable actions in the supply chain.

**What is your recommendation to other procurement organizations about getting started with their own sustainability plan?**
Stop reading this book and get started today! Another tip is to read up on the SDG Ambition Guide and the standards of GRI, and not just the ones that relate to procurement, but all of them. I also recommend that you talk to your CEO and CFO to get commitment and support for your journey.

The next step is to conduct a materiality assessment based on the SDGs to identify clear goals and possible conflicts of interest with other goals within your organization that would need to be addressed.

Finally, communicate, communicate, communicate to get buy-in and increase awareness.

## 8.7 Simon Juul Toft, Lead Specialist: Green Procurement, Ørsted

When asking the suppliers to go the extra mile, a collaborative approach is needed.

## About Ørsted

Ørsted, the world's most sustainable energy company, is recognized as a global leader on climate action and takes tangible action to create a world that runs entirely on green energy.

## About Simon Juul Toft

Simon is a Lead Specialist for Green Procurement at Ørsted and responsible for facilitating the decarbonization of Ørsted's supply chain (Scope 3) through supplier collaboration, innovation, and development. This is a role within the Ørsted procurement department that is tightly linked to the supplier relationship owners of the most strategic suppliers.

## Sustainability in action

**How is the sustainability work structured within Ørsted and procurement?**
The main responsibility for the sustainability work within Ørsted lies with the Global Sustainability Department. Responsibility for supply chain decarbonization lies with the sustainability, HSE and procurement teams in unison whereas the actual execution of supplier decarbonization (i.e., meetings with the suppliers) lies with HSE and procurement.

**We can see that decarbonization has a clear focus in your strategy, could you tell us a bit more about your considerations for this?**
Ørsted has undergone a green transformation in the last decade. In the late 2000s, we were one of the most coal-intensive power generators in Europe with an expanding oil and gas production business. But we took a strategic decision to become a green energy company. To drive the transformation, we invested heavily in renewable energy, particularly offshore wind; exited our fossil fuel businesses, and formulated our vision of "creating a world that runs entirely on green energy." As such, sustainability and decarbonization have been at the core of this transformation and the development of the company as it looks today. We are now well on track to become carbon neutral in our energy generation and operations by 2025, so we are now focusing on the next frontier in our decarbonization journey, which is our supply chain. We aim to be carbon neutral across our entire footprint, including our supply chain, by 2040. Furthermore, we have science based targets in place to guide our journey towards 2040.

**What targets do you have within procurement to support Ørsted's goal of becoming carbon neutral by 2040 for the entire supply chain?**
Procurement's role is to focus on the Scope 3 impact, which includes the impact from the supply chain. For procurement, this means actively engaging with the

suppliers to ensure that they have roadmaps and plans in place that will lead to this goal, and that the category strategies also reflect this ambition.

**Is sustainability a factor that is part of your category management work?**
Decarbonization is integrated into the strategies of the different categories in Ørsted, as a means to steer the overall work and strategic direction. Our category managers are also starting to add other sustainability topics into their strategies.

**How do you engage with your suppliers when it comes to sustainability?**
At this point we see a collaborative approach as the best way to motivate our suppliers to join the same decarbonization journey as us. We are currently focusing on our strategic suppliers for the decarbonization initiative. Through these strategic suppliers we also work with our second-tier suppliers for key materials such as steel.

Decarbonizing the supply chain of renewable energy is complex and requires a significant amount of innovation and collaboration within the industry and across sectors as it is also tied to the decarbonization of other key sectors, including heavy manufacturing and heavy transportation. As such, the core of Ørsted's decarbonization program since its launch in 2020 is to promote collaboration across our base of strategic suppliers in the most $CO_2$ intensive categories towards our common goal. To succeed in this goal, all key actors in the industry must play their part.

**How do you categorize and evaluate the supplier base with regard to sustainability?**
For decarbonization, our initial focus is on the suppliers that we have identified as strategic for Ørsted. For the evaluation, standardized measures are used to make it as easy as possible for the suppliers and avoid it becoming an administrative burden. We make use of the Carbon Disclosure Project (CDP) as a reporting tool. This means that the suppliers only need to answer the standardized questions once and can then share it with as many of their customers as they want.

**How do you support and develop your suppliers?**
We use a structure of continuous and regular collaboration meetings, where we support our suppliers by sharing best practices and suggestions based on our observations and experiences.

**Do you use any tool solutions to help you keep track of the supply chain with regard to sustainability topics?**
We have a Responsible Business Partner Program that uses a tool developed internally, to track our supply chain with regard to social responsibilities.

**Do you have issues with your supplier base not having targets as ambitious as yours?**
We have reached a point where a large majority of our strategic suppliers are interested in supporting our ambitions. For most suppliers, it is evident that this

journey makes sense, and is the right thing to do. Furthermore, we see that suppliers often address this topic as a competitive differentiator. Most suppliers have therefore welcomed Ørsted's efforts and ambition to decarbonize the supply chain. However, we see that suppliers are at different levels of maturity when it comes to developing their own climate strategies.

**What current issues are you facing with regard to sustainability and how will they be addressed?**

The categories where we see the biggest challenges at the moment are steel and high-impact metals from high-risk countries. The risks in these supply chains are addressed by not only being active on a tier 1 level but going further into the supply chain and actively engaging sub-suppliers.

Another challenge is that we want to decarbonize our supply chain by 2040, 10 years ahead of what science requires. We are using a collaborative approach to drive this long-term change and to ensure that suppliers see decarbonization as a competitive differentiator.

**Where do you see supply chain sustainability in 2030?**

The relatability to some of the sustainability topics will certainly continue to grow within the next 10 years. Social aspects, such as child labor or fair working conditions, have always had a high degree of relatability whereas the environmental aspects have been more difficult to relate to. This is however changing as more and more people are witnessing for themselves the direct effects of climate change.

## 8.8 Henrik Larsen, Chief Procurement Officer, A. P. Møller-Mærsk

> We look at sustainability criteria from a TVO (Total Value of Ownership) perspective—if the positive advantages outweigh the increased price, then it's a 'good cost'.

### About A. P. Møller-Mærsk

A. P. Møller-Mærsk is an integrated container logistics company and member of the A.P. Møller Group. With a dedicated team of over 80,000 people, they enable global trade for a growing world.

### About Henrik Larsen

Henrik Larsen started his career as a shipping trainee in Mærsk Line in Copenhagen, Denmark, in 1981. Later, various postings followed in San Francisco, USA, and later in Dubai as country manager for UAE and OMAN. In 2002, he returned to Europe, where he continued his career in Hamburg, Germany, as Line Manager for Germany and Central Europe. In 2003, he began his career in Mærsk Procurement, where he was responsible for various functions before he was appointed Chief Procurement Officer in 2016.

## Sustainability in action

**Mærsk is very active when it comes to sustainability, with plenty of global initiatives and new teams and roles for sustainability (e.g., the decarbonization team). Was there a specific defining moment when it was decided to put sustainability on top of Mærsk's agenda?**
Sustainability has always been on our agenda, but in recent years we began to integrate it into day-to-day business activities and involved local managers more rather than managing things from headquarters.

We have made our voice heard at a UN conference and in the International Maritime Organization assembly to call for joint action and level playing field when it comes to sustainability efforts since it requires massive investments to make the switch. The first step was the shift to scrubbers, which is a system to clean a large part of the emissions while still using regular oil-based fuel. This is something that can be retrofitted onto existing vessels. We are now at the next stage where we need to take more drastic actions and since we want to spearhead the sustainability transition within the shipping industry we have invested in eight new vessels that can be run on cleanly-made methanol. We have done this despite the fact that this fuel is not yet available, but we hope that our investment leads the way and will trigger actions to make the fuel available. We see this as a decision for the greater good—it is bigger than us at Mærsk, this is a decision for the industry.

The same reasoning applies to our decision to establish a research center, the Mærsk Mc-Kinney Møller Center for Zero Carbon Shipping. We are doing this together with partners from the industry and academic institutions to ensure that these topics get the attention they deserve while at the same time finding practical solutions since they are jointly developed with input from the business.

**You evidently apply a holistic approach to sustainability with initiatives to improve the environmental, social, and governance parts. However, there seems to be a strong focus on decarbonization. Is this correct, and if so, is this the main focus now because you see that the other areas of sustainability already have enough attention or that you are already mature in those areas (e.g., anti-corruption, human rights, etc.)?**
Not necessarily. We are in the middle of a strategic review of our sustainability approach to determine where we are, and this goes for all parts of the ESG. This review will also help to determine our ambition level and to set our targets for the coming years. This is being done for the entire company, including procurement of course.

For the environmental, social, and governance parts, our ambition level goes beyond being compliant. It might look like we put a lot of focus on the environmental part since that is what one can read about in the media, but the social and governance parts remain crucial for our way of working. In the industry and regions where we operate these have always been important topics and we will continue to focus on them as much as the environmental part.

**What is procurement's role and how is procurement structured when it comes to sustainability?**
Procurement's main role is the link with the suppliers and making sure that we have the right suppliers that are properly vetted through the onboarding process. We are responsible for demonstrating to our customers that we are compliant and that we offer something more than our competitors.

Do we have complete transparency yet? No, but this is our ambition, and it needs to be our ambition. In the logistics industry we buy what we sell so we need to have a sustainable process and sustainable partners in the supply chain, especially since we want to be the market leader within our industry.

**Does responsibility for sustainability practices lie mainly with a dedicated procurement team or is it integrated within other procurement roles in as well?**
The responsible procurement team mainly has the responsibility for CSR-related topics, such as ensuring that our suppliers and ourselves remain compliant with regard to working conditions, human rights, and other criteria. The overall sustainability work is driven by our Group Sustainability team and we are now investigating if a separate procurement sustainability competence center. In addition, we integrate sustainability practices into our overall procurement processes, meaning that each procurement role has responsibilities when it comes to sustainability.

**What was the trigger behind moving the responsible procurement team completely outside of the HQ to Mumbai?**
We always had this team as a joint setup between our offices in Copenhagen and Mumbai. We found, however, that we had all the right competences needed to run this team in Mumbai and therefore made the switch. To me this should be location agnostic.

**What are your targets within procurement to support the company's overall sustainability goals?**
We do not yet have specific procurement targets for sustainability but we were working on that at the moment as part of our revised strategy. What we do know is that we want to have procurement-specific improvement measurements that can show our progress.

**We can see that you mention digitalization as a driving factor behind your sustainability change and the offering you have towards your customers (through, e.g., TradeLens and Twill). Do you also use any tool solutions to help you keep track of the supply chain with regard to sustainability topics?**
Digitalization plays a big role in our sustainability work when it comes to retrieving the information that we need from the supply chain. There are a lot of solutions out there and we are currently running several pilots to determine the best way to gather data, how to analyze it, how to get the transparency that is so important, and then to ensure that everything is properly documented through the entire supply chain and onwards to our customers.

**Is sustainability part of your category management work?**
We are considering how to build sustainability within our category strategies, both to ensure compliance but as mentioned we want to go beyond that, so it is also about finding the best sustainability offering within each category.

**Is sustainability part of your decision-making process?**
We have started this process and are aiming for sustainability to become a qualifier from our suppliers. The hope is that it eventually becomes an add-on since we are selling this service to our customers.

**Have you had to balance sustainability against other traditional decision criteria, or did this come naturally?**
Criteria such as price will always be relevant. The way that we look at it though is "good cost" and "bad cost." If there is something that costs more because it is driving the sustainability agenda forward, then we see this as a "good cost." It does not mean that we are not discussing prices and that we are not adding arguments to our negotiations, but we look at it from a TCO (Total Cost of Ownership) and TVO (Total Value of Ownership) perspectives and calculate what possible positive advantages could come from the extra costs.

**How do you support and improve the suppliers on their sustainability journeys?**
We are working together with our suppliers to provide them with what we have learned from our sustainability journey. We do not always have "the golden answer" but where we have gathered knowledge and experience, we share it with our partners in the supply chain so that the whole ecosystem around a certain process can improve. We see that as an important part of our role within the supply chain and it goes both ways—we are also happy to learn from our suppliers about how we can improve our own practices. These learnings do not only come from our suppliers. We learn equally as much from other partners in our ecosystem, such as partners in our procurement network of non-competing industries.

**What do you see as the biggest challenge when it comes to fully integrating sustainability within procurement at Mærsk?**
One major challenge is how we can be sure that we really get what we need. We are reliant on external parties, tool providers, and our partners in the supply chain to provide us with information and data, but how can we be 100% sure that it is correct? The amount of time and resources that you need to invest also require careful consideration and planning.

**Where do you see the topic of sustainability in 2030?**
My hope is that it will be something that we talk less about because it will already be such a natural part of how we work and do business. This would mean that we have reached the tipping point where we are finally starting to go in the right direction globally.

**What would speed things up?**
We need joint agreements to speed up progress on sustainability. We are coming towards that now and there are fewer and fewer players in our industry that are "taking the easy way out." The majority embrace sustainability and are willing to work towards a better future. No one can solve this alone.

## 8.9 Chaojun Li, Global Head of Sustainability, WS Audiology

> Sustainability is one of the three drivers to deliver our ambition to become a clear industry leader.

### About WS Audiology

WS Audiology develops, manufactures, sells, and distributes hearing aids. WS Audiology's ambition as a global leader is to unlock human potential by making wonderful sound part of everyone's life. wonderful.

### About Chaojun Li

Chaojun is currently the Global Head of Sustainability at WS Audiology. Her experience of sustainability stretches over more than a decade. Before WS Audiology, she was a sustainability advisor at both the energy company Ørsted and the logistics company Mærsk.

## Sustainability in Action

**Can you describe what your role as Global Head of Sustainability at WS Audiology involves?**
My team and I are responsible for driving sustainability within the organization as well as the value chain. We have two focus areas: integration and communication.

Integration—Sustainability should be integrated into the operating model rather than something that comes "on top," which means that all departments need to be closely involved and take ownership of the various sustainability programs.

Communication—It is also crucial that we communicate our sustainability efforts and performance to the relevant stakeholders and the general public. We communicate via our sustainability report, our website, social media, etc.

**How is your team set up and how is it anchored within the organization?**
We have three full-time staff and one working student. One person is in Singapore and the rest in Denmark. I report to the chief quality and regulatory affairs officer who reports to the CEO.

**How do you ensure the anchoring of the strategy within the organization?**
We started with a materiality assessment that involved discussions at Group management level from different functions to determine which topics that are important to WSA. Once we had decided on these topics, we then aligned with the team members one level below management and assigned program managers for specific initiatives. The teams below them in turn assigned activity managers who manage part of an initiative. We in the sustainability team are the overall owners of all initiatives—sustainability is one of the three drivers to deliver our ambition to become a clear industry leader.

The materiality assessment has also been passed on to five of our key suppliers since we depend on our supply chain to support our targets and to have similar ambitions themselves.

**You have a decade of experience in different sustainability roles, what do you see has changed in relation to sustainable procurement?**
What has changed dramatically is the general focus on sustainability. When we started to introduce sustainability in the supply chain 10 years ago it was still new to many suppliers, and also to the internal procurement managers. However, with the enforcement of regulations, along with pushes from consumer and investor, sustainable supply chain becomes a must-have for all companies. Moreover, companies start to collaborate with suppliers on sustainability topics to find solutions to tackle pressing issues and go beyond just compliance.

**When did WS Audiology decide to get serious about sustainability? Was there a defining moment?**
Sustainability has always been in the DNA of WSA and also the two companies that merged to become WSA. When our CEO, Eric Bernard, came on board in 2019,

decision is made that we want to manage sustainability in a more systematical way and disclose more what we do to the public.

**You state that you are currently contributing to five of the 17 SDGs. Do you see a risk in "cherry-picking" what you contribute to rather than taking a holistic view?**
We are focusing our attention where we can make a real impact. We base our reporting on GRI and SASB standards, which give us a very detailed reporting method for how we influence the company and society. While SDGs are designed at national level, an individual company can make a meaningful contribution to some of them.

**How is the procurement team involved in sustainability?**
Our procurement department take the lead in sustainable supply chain management. They make sure the Code of Conduct for Suppliers is signed by relevant suppliers. For high-risk suppliers, supplier Code of Conduct audits are conducted by third-party auditors.

We in the sustainability team work closely with procurement on supply chain decarbonization.

**How is sustainability integrated into procurement other than the specific initiatives such as decarbonization?**
We are working on systematically incorporating sustainability into the requirements for our suppliers. As an example, since 2021, we included sustainability requirements to suppliers that provide dummies. The product dummies are really our ambassadors. They are used in every store where our hearing aids are sold, and we ship around 1 million each year. We worked with our suppliers to make the dummies and the packaging of them more sustainable, including using recycled materials and reduced packaging.

**How do you categorize and evaluate the supplier base with regard to sustainability?**
We categorize suppliers based on spend, category, country, and type of product or service to identify those that are the highest risk. We then audit the high-risk suppliers every 2 years. We have conflict minerals and other high-risk metals in our supply chain and for these our suppliers need to inform us where the minerals and metals came from. In addition to standard audits, we have a process called Product-Related Environmental Protection (PREP) to ensure that there are no forbidden materials in the products, and that restricted materials stay within assigned limits. This special auditing process is run directly by the sustainability team.

**Where do you see sustainability in 2030?**
I believe that climate change will remain highly important over the next decade. By solving the challenge of climate change we will also cover many of the other

challenges related to environment, such as air pollution, overconsumption, deforestation, and biodiversity.

Human rights and especially human rights issues in the supply chain will also keep gaining traction.

## 8.10 Dr. María Mendiluce, Chief Executive Officer, We Mean Business Coalition

Procurement leaders need to be at the forefront in the struggle to reduce scope 3 emissions.

### About the We Mean Business Coalition

We are a global non-profit Coalition working with the world's most influential businesses to take action on climate change. Together, we catalyze business and policy action to halve global emissions by 2030 in line with a 1.5 °C pathway.

We collaborate with a core group of seven business-focused climate non-profit organizations to accelerate the transition to a just and climate resilient net-zero economy. These founding partners help to deliver key initiatives and activities that are central to the success of the overall Coalition's bold objectives. Beyond this core group, we also collaborate with a wider network of organizations that help to realize our shared vision at speed and scale.

### About Dr. María Mendiluce

María Mendiluce is the CEO of the We Mean Business Coalition and brings 25 years of experience working in business, international organizations, academia, and governments in sustainable development, energy, and climate action.

As chief executive of the Coalition, María is working to mainstream net zero in the business sector encompassing companies of all sizes and sectors. She sits on the Executive Board of The Science Based Targets initiative and has taken an active role in the founding of the Mission Possible Partnership, an initiative that develops decarbonization roadmaps for the high emitting sectors. María also co-founded the SME Climate Hub, a one-stop-shop for smaller companies to commit to net zero and work with the value chain on implementation.

Formerly Managing Director and Senior Management Team at WBCSD, María has a history of establishing transformative industry projects. These include the Alliance to End Plastic Waste which pledged $1.5 billion to address plastic leakage to the environment; the Transforming Urban Mobility project to drive clean, safe, and efficient mobility for all and WBCSD's flagship Low Carbon Technology Partnership initiative which was recognized as a key contributor to the Paris Agreement.

She has also held senior positions at the Economic Bureau of the Spanish Prime Minister, in the CEO's office of Iberdrola and at the International Energy Agency. María has a Ph.D. in energy economics. She is a Commissioner on the Energy Transitions Commission, a Co-Chair of the Global Environment Outlook Business Advisory Board (UNEP), and she is a board member of the Centre for International Environmental Studies (CIES) in Geneva.

María was awarded the Spanish Climate Personality of the year in 2021 by The Climate Reality Project in Spain as part of the third edition of their Climate Leaders Awards. María has been listed as critical influencer on climate action by different media groups such as GreenBiz or Eddie. María is Spanish, has three kids, and lives in Geneva.

## Sustainability in Action

**Why do you think some countries are still subsidizing, or even worse, still investing in fossil fuels, especially coal?**

I think we need to distinguish between the different fossil fuels. There are not many subsidies for coal, but some countries are still investing in it—the International Energy Agency notes that over the next 5 years more than 300 new coal-fired power plants are set to come online, mostly in emerging or developing economies. And coal remains deeply embedded in the power systems of many countries—for example, coal accounts for more than 60% of electricity generation in India, China, and Indonesia. However, many other countries have said they will stop using it and substitute it with renewables or gas.

At COP26, governments emphasized the importance of phasing down coal and removing fossil fuel subsidies, along with landmark agreements on the phasing out of coal, which shows real progress is underway. However, in order to be in line with the 1.5 °C pathway, coal-fired power generation must be phased out by 2030 for advanced economies, and by 2040 for other countries. Business expects coal-fired power to be phased out in order to keep the 1.5 °C target within reach and is investing in the clean energy transition.

I think it is interesting to see how China is doing given it is the world's largest energy consumer. There is still high demand for coal domestically, for its power and industrial sectors, and limited natural gas alternatives there. But China is also working on renewables and they are fast. Over the next 5 years, the IEA forecasts that China will increase its renewable electricity capacity by close to 800 GW, which for context, is roughly equivalent to the existing installed global capacity of solar PV (739 GW).

Renewables are more competitive than coal so we hope market forces will win out, but it needs time to make that transition happen. Coal has no place in the world.

Oil's place in the world is also diminishing as the uptake of electric vehicles accelerates. There will still be some used in aviation and shipping but eventually there will be alternative fuels in these sectors as well.

Although gas produces half of the emissions of oil it still produces emissions. Because of the current high energy prices people think they need to invest more in gas production, but we should actually be investing more in gas storage because the same storage facilities could be used for hydrogen in the future. Natural gas pipelines could also be repurposed to help transport hydrogen.

The future is renewables and green hydrogen, and we need an infrastructure that supports them.

Interest in green hydrogen is growing given its versatility, but it is not yet cost-competitive. We need massive investment in infrastructure and innovation before it will be possible to transition to a world where hydrogen plays a major role. This includes investment in storage and transport as well as the ramping up of electrolyzer manufacturing capacity and the expansion of power infrastructure to meet demand. This should be the focus of governments.

**The G20 represents approximately 90% of global GDP and almost 80% of global trade and greenhouse gas emissions. Should not these countries be cleaning up their acts and setting an example for the rest of the world to follow?**

It is important to make a distinction here. There is agreement and consensus within the G7 on emissions targets, this is important given that the G7 accounts for a quarter of global energy related $CO_2$ emissions. However, within the G20 China and India are still using a lot of coal. So, there is push-back.

Domestically, coal is causing a lot of health problems in China at the same time as the supply chain is coming under pressure. As a major exporter, this is also a problem for China. Companies want to have sustainable supply chains and $CO_2$ is an increasingly important part of the procurement decisions. If China does not move these companies might switch to markets that can move. China needs to realize that investing in coal no longer makes sense. We are not there yet but we are going in the right direction.

**A coalition of 778 companies, with revenue equivalent to the GDP of the UK, are calling on the G20 to end financing and development of new coal power plants and to phase out coal-fired power generation by 2030 for advanced economies, and by 2040 for other countries. Do they have the leverage to make this happens?**

The G20 leaders have heard the clear calls from business on financing coal and providing climate finance and chosen to act to keep 1.5C within reach. Building from the G7 outcomes earlier this year, the whole G20 has now committed to end international coal financing by the end of 2021. G20 leaders have recognized the need to transform their energy systems in line with the Paris Agreement and agreed to help each other to do this though they stopped short of putting an explicit timeline on phasing out coal. Now this momentum needs to be backed up with action.

Many companies and countries have ambitious plans and have committed to net zero, but this requires a massive transformation that we are just starting now. When there are lots of companies in this transformation it makes it much easier because a significant volume of emissions come from the supply chain and many companies have shared supply chains. So, a unified movement is a positive start because it will encourage everybody to achieve their targets. So, the question is: How can you mobilize companies and countries to be ambitious and take action? Tracking progress is now very important because when investors and ESG professionals look at who is performing better we have good metrics to track them. Then there is competition on, and this will accelerate the transformation to a zero-carbon economy.

One thing is very clear to me: in most cases, when it comes to companies publicizing their commitments and tracking their progress, they tend to outperform their own targets. We need much more of that. Unfortunately, the opposite is true of some countries. But if we could get countries working closely with their domestic companies to achieve their targets, we could move from a system that is voluntary to one where it is mandatory for businesses to be net zero and reduce emissions. We are not ready yet but in a few years, because things are getting tough with climate change, we will get there. This means that it is better for companies to get ahead of the game now before they are forced to later.

**Is a carbon cap and price mechanism going to be the enabler we need? Are you pushing for that in your discussions?**

I think carbon pricing is fantastic. I think renewables are competitive even without carbon pricing, but we need carbon pricing as it can help steer the type of investments made today, with implications for future emissions. We need to be efficient with how we use the energy even if the energy sources might be cheap. Because the resources are finite. A well-designed carbon price can provide some certainty to trigger investment in low carbon technologies or longer-lived/longer horizon infrastructure and RD&D (such as Carbon Capture, Utilization and Storage (CCUS), hydrogen, etc.).

Companies should include an internal carbon price that reflects its true costs. That will help them to make decisions that are sound. And it provides an internal price signal, so they do not make investments that are carbon intensive. A very interesting example is that cement companies are now included in the EU's Emissions Trading Scheme.[3] And then suddenly, wow so many innovations! The industry made commitments and now say they can reduce emissions by 30–40% by 2030. This is something that would never have happened if there was not a financial incentive to cut back on emissions.

**Do you think the carbon pricing will become an established part of the economy?**

Carbon pricing is going to come in patchworks. But I think it has proven to be a very good instrument for companies to reduce their emissions and reduce prices. So yes, I definitely think that it will become a standard feature of the economy.

[3]EU Emissions Trading System: the world's first and biggest carbon market.

I also think that companies, boards, and CEOs and management teams will have $CO_2$ targets as part of their incentive and compensation schemes. This will be cascaded down through the organization so that procurement will have both an environmental cost and a financial incentive and they will optimize both. Once carbon pricing becomes part of the objectives of companies, they have to include it as part of their procurement strategy.

We are not there yet, but companies are going to reduce scope 1 and scope 2 emissions rapidly but they will struggle to reduce scope 3 emissions. Then they will look at procurement offices and ask, "what are you going to do about this?" Procurement leaders need to be at the forefront in the struggle to reduce scope 3 emissions—they need to know that this is coming so they should already get started.

**How can procurement organizations and decision-making processes in companies support the race to zero? Are there any other recommendations you have for procurement and sustainability teams to help speed up decarbonization?**

Between 50% and 95% of the emissions of large companies come from the supply chain. More than half of them come from the energy-intensive sectors—materials such as steel, cement, and transport such as trucking, aviation, shipping. Countries need to do something about this because their footprints depend on those companies making progress with their emissions. A solution could be the creation of demand clubs and there are some nice examples emerging, such as SteelZero.

Then there are the small and medium-sized enterprises that we have to think about. There, two things are needed. First, SMEs must be educated about the need to adopt sustainable business practices because of the benefits this will bring to them. They do not necessarily have the resources or time to do this themselves, so they have to be helped by the bigger companies they are supplying.

After this, sustainability needs to become a condition of doing business. Procurement can start setting up specific contractual requirements, such as disclosing sustainability-related data, that helps SMEs understand that sustainable business practices are manageable and that they can actually lead to cost savings, that sustainability can be a great opportunity. You start with a carrot and then you end with a stick. Then at some point, once you have around 20%–30% of your suppliers are on board, then you can make those things mandatory so that the rest will follow. And the rest could leapfrog because they can learn what other companies in the supply chain have done right.

**So that is only your suppliers but then on how do you convince the big industries of the need to come together? For example, to meet the net-zero targets, the automotive industry needs zero emissions steel.**

Industry needs to find a way to join forces and motivate suppliers to invest a large amount of money in zero emissions steel that will spread the premium over multiple players and make the investment much more attractive for everyone. In the end, collaboration like this might be less expensive than offsetting or paying carbon penalties in the future.

I think procurement has an enormous responsibility to explore innovative ideas that will bring benefits to the company. Right now, the benefit is not very clear because there is no carbon pricing included. Ultimately, consumers are asking for reduced $CO_2$ footprints on products, so you will have to trickle this down to the supply chain and make it happen. That is the future.

## 8.11 Markus Strangmüller, Corporate Development and Sustainability Manager, Siemens

Overall, we see sustainability as a business opportunity [...] it's part of our strategy discussions.

### About Siemens

Siemens is a technology company focused on industry, infrastructure, mobility, and healthcare, creating technologies and transforming the industries that form the backbone of economies.

### About Markus Strangmueller

Markus Strangmueller leads the Sustainability Strategy and Governance department within Siemens AG. He regards himself as a "Business Humanizer" and stands for purpose, trust, and legacy.

### Sustainability in Action

**What is your interest in sustainability?**
Nature plays a huge role in my life. I have two children, which is why the next generation means a great deal to me. Overall, I believe it is important to get the right balance between "me" and "we," between short term and long term, and between people, planet, and profit.

**You have been responsible for sustainability at Siemens for more than 10 years already. How would you describe the journey and are there any notable milestones?**
If you look at Siemens as a company, it has been very strongly characterized by sustainable and long-term development from the very beginning. Social responsibility was an important topic in the days of our founder, Werner von Siemens. In 1971,

we set up a department to look at environmental issues but over time social responsibility became the priority. In 2008, we started looking at supply chain management and sustainability and then in 2009, we opened a dedicated sustainability office. At that time, we defined and built a strong environmental portfolio. We set up a program with three pillars. Business opportunity is the first pillar because we have the advantage that our portfolio elements already contribute very broadly to the sustainable development. The second pillar is that if I want to address business opportunities, I naturally need to walk the talk. The third pillar was stakeholder engagement. We cannot do many of these things on our own, but together with others. The question then arises as to which topics we should work on with which stakeholders. There have been many different facets in various forms. We have always taken a broad view of all aspects of sustainability.

**Many companies still view sustainability from a corporate point of view. Which responsibility or contribution do you see with the business units, divisions, or regions?**
A good combination is needed because for certain topics you basically need a corporate decision. There are always discussions about how to make funding available for certain topics. But there are also discussions about how to integrate sustainability-related opportunities and risks as part of a product lifecycle management (PLM) process. This cannot be done centrally but must be done within the individual business unit.

**To what extent is cross-functional cooperation about sustainability taking place?**
At the moment, decarbonization is a very big issue, which is why we have defined quantitative targets that include targets for scope 1 and 2, as well as scope 3, including scope 3 upstream. But how do we achieve them? We collaborate with our suppliers so we are transparent and then see what can be done to reduce $CO_2$. One solution might be for a certain supplier to say that they could offer different materials. Our buyer has to talk to the PLM colleague to see how this would affect the design and influence the product.

Collaboration happens precisely at the process interfaces. But it is not a one-way street. We approach the PLM community and ask them to what extent they consider sustainability-related opportunities and risks in their process. It is not about defining every milestone, it is also about a change of mindset—we want to get our design and product teams actively thinking about sustainability and circularity within the framework of a journey. This includes potentially using different materials and that is where there is strong collaboration with SCM.

**Are there any regular meetings between the SCM colleagues and procurement?**
Each relevant unit has a sustainability manager. The sustainability managers are then in a community that we orchestrate. There are meetings between the governance owners only, or with only the countries or the businesses. On top of that, there are

topic-related, cross-functional sessions. Then, every 2 years, we usually have a sustainability manager summit.

**How do you align and integrate the business units to contribute to the corporate goal of decarbonization or circularity?**
The corporate targets have been given to the units. Supply chain management has responsibility for reducing upstream scope three $CO_2$ emissions. They are responsible for driving this target forward with suppliers and then driving it forward again with the individual business units and their procurement departments. Then we focus on changing attitudes to sustainability at corporate level. We have now established internal control requirements for governance where the CEOs are responsible for ensuring that sustainability-related opportunities and risks are implemented strategically and systematically in their business processes and decisions. What we then do is to tell them that there are a handful of questions in the SCM, PLM, or CRM processes that should be considered. Whether it is through milestones, whether it is through dialogs, or whether it is in the escalation process.

**Mindset change is an important part of any transformation. Can you give me an example of how you approach this at Siemens?**
For example, we organize workshops with PLM communities where the focus is on the questions of how sustainability and circular design can be built into the products. We also have a range of training and learning programs about the circular economy from basic to expert to advanced. Education around the topic is important, but then there is engagement followed by the practicalities of implementation. Education alone is not enough.

**When it comes to procurement, are people told that they just have to implement sustainable practices or is there a dedicated staff position?**
We have basically three different blocks. The first one is the governance owners, the next is the businesses, and the last is the countries. Governance owners must define policies, establish management systems, set targets, have KPIs, define long-term goals, and look at budgeting. They are responsible for the control behind it, for the disclosure in their area. And that is again across businesses and across countries. For sustainability, we have a sustainability board with a sustainability office above it for the group. The task there is to specify the strategic direction of sustainability and to then operationalize it. The implementation then takes place in the businesses and in the countries.

**Who is part of the Sustainability board and how strong are these two committees?**
Our Sustainability Board, which is the second-highest committee we have in the Group, is chaired by Judith Wiese as Chief Sustainability Officer. The CEOs of the businesses are represented, so Smart Infrastructure, Digital Industries, Mobility, and our Portfolio Companies. Then there are four country representatives from the most important countries, but also from other countries where other requirements come

into play. Additionally, there are the heads of the corporate functions when relevant if they are not already covered by other representatives. That is the Sustainability Board, the core committee for the strategic orientation regarding sustainability. They meet four times a year and discuss these topics. Procurement is represented indirectly but attends the Sustainability Board at least once a year when special topics such as decarbonization are discussed.

**How many sustainability managers are there in your organization?**
Probably about 60–70 people who are nominated. In the SCM department, there is a separate sustainability unit that takes care of supply chain due diligence law and decarbonization of the supply chain, for example. They are a separate team that defines the policies for the supply chain. They deal with supplier audits and how audits are carried out, what kind of approach is used, and so on.

**But that would also mean that the development departments in the individual divisions would then all have to make their contribution accordingly.**
Change accordingly, of course. In a policy board like this, there is a central procurement department and then there are procurement departments in each business, in our case the Digital industries and Smart Infrastructure and Mobility. If they have issues, then of course they talk it through again with their function owners in the areas.

**How are you getting prepared for the Supply Chain Act?**
We have a task force made up from our legal department and people responsible for the supply chain and they are defining how the act will affect Siemens.

With the supply chain due diligence law, there are several issues that have to be taken into account. You also need a mindset that is aware of what is going on and why this is the case. Failing to keep up with the times is extremely risky and could threaten the existence of your business. There are more and more requirements, and the global community has agreed on the 2030 agenda but so far there is limited progress. That is why the EU is really pushing the pace on all the legislative issues because they see that if we do not step on the pedal now, we will never reach the 2030 targets.

**What specific role does procurement play?**
For me, procurement is one of many roles that plays a part in sustainability. We have two core targets regarding the supply chain: first, an ESG secured supply chain, and second, $CO_2$ reduction within the supply chain. These are essential to our long-term objectives. But decarbonization of our own operations is also important. Overall, we see sustainability as a business opportunity so it is part of our strategy discussions. It is also an important consideration with the PLM or the product designers because they aim to develop products that help society to decarbonize or move into the circular economy.

So, procurement has an important role to play in sustainability, but so do many other functions.

**How does Siemens create transparency about scope 3 emissions data? Do you have your own system?**

We have established a database with an external partner. From this we can derive which $CO_2$ emissions come from which material group and from which country. That was the first step to achieving $CO_2$ transparency. We buy steel from different countries and how does it fit together, what are the energy mixes there and how do we get that right? Additionally, 2 or 3 years ago we wrote to the 8000 or 9000 of the largest $CO_2$ emitting suppliers and had discussions with them about the extent to which they are already dealing with the issue and what measures they are taking. We created awareness in this first step.

We have now developed a carbon web assessment tool, which is being rolled to the various suppliers, in order to increase transparency and thus also to bring more actual data into the tooling so it can be adjusted. When the transparency is there, the next step is to have discussions around $CO_2$ emissions in the whole value chain so that we not only get a cost but also a carbon footprint for the products.

We have another project running in parallel internally right now to increase transparency in the supply chain. There are initial pilots for this in various areas. But that is now the journey that has been started. We said that we want to be $CO_2$ neutral in the supply chain by 2050 and we want to reduce 20% by 2030. But that is not just at the T1 level, you also have to go to T2 and T3 and talk to them again about what they can do there.

**Which other topics do you see coming up besides decarbonization?**

We have already included the topic of net-zero operations in our long-term incentive (LTI), which we announced as a target back in 2015. I would not be surprised if Scope 3 upstream was to be included in the near future, in 1- or 2-years' time.

We have made a strong start on the topic of ESG risks in general. In 2007, 2008, we introduced our code of conduct, which is also being successively developed. Operating globally, we have to comply with the United Nations Guiding Principles on Business and Human Rights and as well with the UN Global Compact. The next level is the business models of these ranking agencies that are now emerging. In large companies like ours, for example, a large percentage of our suppliers are also our customers on the other side. That is why, we have to look at how balanced we are going forward. This is a journey and not something we can solve overnight.

**What would be your recommendations for a procurement officer?**

The one issue we cannot avoid today is having a code of conduct. This at least covers the minimum requirements.

The second topic is certainly to go somewhere in the direction of $CO_2$ transparency because that will simply be indispensable. As a procurement officer, I would need to consider whether I need partnerships to address certain issues that I would struggle to address on my own.

If I have a code of conduct, then of course I have to consider what approach I will take to check that it is being adhered to, for example, using a risk-based approach, or using a different approach such as self-assessment or external audits. I probably would not overcomplicate it at the beginning.

**What is best practice for you?**
There are already many in Germany. There are some that are now very active and really progressive in an environmental area. But I think there are already many medium-sized companies in Germany that are just starting out and do not have the resources behind them. But they still think that they have to do something for sustainability on the side. I hope that German SMEs will be able to think in this integrated way, because that is of course essential for the future.

## 8.12 Thomas Udesen, Chief Procurement Officer, Bayer, and Co-founder of SPP (Sustainable Procurement Pledge)

> Our ambition is to have equipped and empowered one million procurement professionals by 2030 so we have a lot of work and little time.

### About Bayer and The Sustainable Procurement Pledge

Bayer is a global enterprise with core competencies in the Life Science fields of health care and agriculture. SPP brings together Procurement professionals to ensure responsible supply chains.

### About Thomas Udesen

Thomas is the Chief Procurement Officer of Bayer and one of the co-founders of the Sustainable Procurement Pledge (SPP). At Bayer he leads the international procurement strategy with a budget of ~24 billion euros, where the guiding principle is sustainability. Aside from Bayer, Thomas Udesen co-founded the Sustainable Procurement Pledge in order to collaborate with other procurement leaders and drive supply chain sustainable change.

### Sustainability in Action

**When did Bayer decide to really step up the efforts on sustainability? Was there a defining moment?**
Bayer's conviction of inclusive growth and responsible business conduct has grown and evolved over several decades. We were one of the first companies to sign the principles of the UN Global Compact in 2000. Bayer was also a founding member of industry collaborations like Together for Sustainability (TfS) and Pharmaceutical

Supply Chain Initiative (PSCI). So, there was no defining moment—the awareness and the commitment are deep-rooted in the company, and the practices have developed and strengthened continuously over a long time.

In 2019, the sustainability strategy for the company was reinforced and given a further lift by the recruitment of Matthias Berninger as SVP for sustainability. This further strengthened the team as he brought a wealth of knowledge from his experiences with Mars and as a parliamentarian for the Green party in Germany.

**Do you believe that the nature of your business meant you had to strengthen your focus on sustainability?**

The focus comes from a strong commitment to sustainability. Bayer is a large company (100,000 employees and more than 50,000 suppliers) so there are occasions when some individuals and companies step outside the desired boundaries, but this is why we have our checks, controls, and our overall management system. Our sustainability practices have made it possible for us to react swiftly to these events and make further improvements.

**How is your procurement department structured?**

In 2016, Bayer went from a decentralized procurement setup to one integrated into a global organization, both for direct and indirect procurement. This meant a complete change of the operating model and the underlying processes for procurement. However, the focus on sustainability remained and we have continued to invest in this area and in innovation and people and culture.

**What targets or goals within procurement support Bayer's sustainability targets?**

Procurement is contributing to all of Bayer's company targets for sustainability. Scope 3 is a focus area, but procurement also contributes to Scope 1 and 2 reductions. We also have targets related to the overall inclusiveness and resilience of our supply chain, specifically how well we manage potential risks. We measure this by both the coverage rate but also on the ratings and improvement in ratings.

Inclusiveness is a big topic in the USA, and it is growing in importance in Europe. It is still at the initial stage in Asia but we are working hard on this so that it becomes equally relevant globally.

To ensure a resilient supply chain we also monitor based on the risk. There are some drugs we produce that are needed for human survival so we simply cannot accept any supply chain disruptions, which is why we monitor our supply chain in real-time down to the n-tier. A resilient supply chain is a sustainable one.

**What systems or tools are you using to track the performance and risks in the supply chain?**

We use EcoVadis for assessing the supply chain with regard to different sustainability parameters. We work with the CDP for tracking $CO_2$ and we monitor our supply chain in real time with DHL360 Resilience.

**How do you engage with your suppliers when it comes to sustainability? It stated in your 2020 report that 670 suppliers had been evaluated—does this mean that the focus is on the most strategic suppliers?**
The primary focus has been on strategic suppliers and those suppliers in either high-risk regions or those that provide high-risk products (e.g., products containing conflict minerals). We are continuously adding to this list, and at the same time we follow up with registered suppliers to ensure that they are making the improvements we agreed on.

**Are there any problems stemming from the fact that your supplier base is not as far advanced as you are when it comes to sustainability?**
Yes, this is a challenge. Many small and medium-sized enterprises that are further away from the end customer in the supply chain have not yet risen to the challenge and implemented sustainable practices and processes. It is important for us to keep stressing that this is not something that only larger corporations have to deal with, it needs to come from all tiers of the supply chain and from all companies regardless of size.

**Is sustainability considered in your category management work? Is it part of your category strategies?**
Sustainability is part of our category management work and consequently part of our category strategies. However, the level of maturity differs from category to category. In some cases the category has a very detailed roadmap for how they will contribute to the sustainability targets. In other cases, for example, when it would require us to completely rethink a product, we might not have the same level of detail.

**Is it part of your sourcing decision process? Have you had to balance sustainability against other traditional criteria, or did this come naturally?**
Sustainability is part of the sourcing decision process, and this comes quite naturally since it is also part of our management targets. In many ways the discussion around sustainability follows the same pattern as the one that took place about quality a few decades ago—we are starting to develop more and more mature ways of calculating the TCO including sustainability, and this will most likely only be stronger in the future. As the cost of $CO_2$ and other sustainability factors increase the business case for sustainable products and services will improve. We do not see a conflict between targets here but rather a complex problem that we are all working together to solve.

Excluding a supplier from the supplier pool for sustainability reasons is something that we can do but prefer not to. When Bayer walks out on a supplier it could mean that the employees of that company end up unemployed and there is nothing that guarantees that the supplier will improve for another customer. That is why we prefer working together with our suppliers to ensure that the requirements are met and that there are continuous improvements.

**Where do you see sustainability in 2030?**
I remain optimistic and believe that we will have solved a lot of the issues we see today. My hope is that sustainability does not become too politicized because it is an issue that affects everyone. Within procurement, we also see room for new business models and collaborations with suppliers, such as joint contributions to sustainability initiatives.

**What are the current sustainability challenges that you are facing and how will they be addressed?**
There is a general challenge to connect the "holy trinity" of innovators, regulators, and the markets. We need innovators to ensure that we find solutions to the sustainability problems where we do not yet have a solution. At the same time, we need a market that is requesting these solutions and regulations that support them, levels the playing field, and ensures that the right behavior is rewarded. What we currently have with these three parts is a system that is broken. The innovators exist but are not always rewarded for solving the right issues. This also goes for procurement where we in some cases could be even more innovative in finding new solutions. The regulators on their side are not taking their role seriously enough and the market is not always ready to fulfill their part of the responsibility.

Behind all of this is a lack of change management. Everyone keeps saying that we need to do things differently and that what has been done in the past is no longer relevant for the future. For a lot of people this is a big undertaking, and most companies and societies are not investing enough in the people aspect of things so that the competence gap is closed.

**Can you briefly say what SPP is (spp.earth)?**
Founded in 2019, the Sustainable Procurement Pledge (SPP) is a pro-bono community of those committed to embedding sustainability into their procurement practices—and sharing the knowledge and tools to do so. Sustainable Procurement Pledge GmbH was registered as a charitable organization in Germany on July 1st, 2021.

**We know from earlier discussions that your work with SPP came out of the 2019 big push for sustainability and the need to democratize knowledge for procurement practitioners to help them contribute. Have you seen a mindset shift since practitioners started to get more support and are now able to show more examples from their sustainability work within their companies?**
Definitely. 75% of the ambassadors are now saying that they are able to make better decisions now.

**You are now moving from organizing the movement to creating chapters and taking things from ideas to action—how will you measure success?**
The current measures for success will remain, so the ambassadors' satisfaction and their ability to make better decisions, and if they have the right knowledge available. Engagement, reach, and behavioral change will also be reviewed. Our ambition is to

have equipped and empowered one million procurement professionals by 2030 so we have a lot of work and little time.

**What advice would you offer a practitioner who wants to start talking about sustainability within their procurement organization?**
Do not get intimidated, do not lose hope, and *build your vocabulary*. We are our conversation—if we succeed to bring our point across in a convincing way with factual arguments the experience is that most managers will be open to listen and to take things further. Also, do not feel that you are alone. We all struggle with the challenge so make sure that you talk to your peers, build alliances, and do not reinvent the wheel. The African saying very much applies: "If you want to go fast, go alone but if you want to go far, go together"!

# Appendix

## Guidelines and standards

| Name | About | Access |
|---|---|---|
| UN Global Compact and UN Sustainable Development Goals (SDGs) | The 17 goals provide a powerful aspiration for improving the world, showing what actions are needed. The multi-year strategy of the UN Global Compact is to drive business awareness and action in support of achieving the SDG's by 2030 | https://sdgs.un.org/goals |
| OECD (Responsible Business Conduct) | The OECD guidelines give both general guiding but do also offer industry-specific advice | https://www.oecd.org/investment/due-diligence-guidance-for-responsible-business-conduct.htm |
| Environmental, social, and governance (ESG) | Environmental, social, and governance (ESG): Set of standards for a company's operations that socially conscious investors use to screen potential investments<br>• Environmental: Contribution to climate change through GHG emissions, waste management, energy efficiency<br>• Social: human rights, labor standards in supply chain, workplace health and safety<br>• Governance: set of rules and principles targeting rights, responsibilities, and expectations between stakeholders | |
| EMAS | Focuses on environmental impacts and the entire product/service lifecycle | https://www.emas.de |
| GRI Standards | Create a common language for organizations to report on sustainability impact in a | https://www.globalreporting.org |

(continued)



S. Aichbauer et al., *Responsible Procurement*, Management for Professionals,
https://doi.org/10.1007/978-3-030-98640-7

| Name | About | Access |
|---|---|---|
| | consistent and credible way. This includes reporting standards for • Scope 1: direct GHG* emissions • Scope 2: location-based energy indirect GHG emissions • Scope 3: other GHG emission | |
| SASB Standards | Enable businesses to identify, manage, and communicate financially-material sustainability information to their investors | https://www.cdsb.net/sustainability-accounting-standards-board-sasb |
| Task Force on Climate | Related Financial Disclosures (TCFD)—is recommended and supported to enhance more informed investment, credit, and insurance decisions | https://www.unepfi.org/climate-change/tcfd/ |

Alliances and Initiatives:

**Enhance networks and collaboration**

The challenges posed by sustainability represent a massive undertaking but there is no reason to do it alone. Collaboration with NGOs or intergovernmental institutions, industry-specific initiatives, suppliers, employees, and even activist groups can create synergies that could accelerate the transition to sustainable business. The following tables give an overview of some of the sustainability guidelines and standards that are available, industry alliances and initiatives, and sustainability networking platforms.

| Category | Sub-category | About | Access |
|---|---|---|---|
| Industry-specific initiatives | Automotive | *Drive Sustainability* is a partnership between global automotive companies that aims to drive sustainability throughout the automotive supply chain | https://www.drivesustainability.org/ |
| | Chemicals | *Together for sustainability (TFS)*, a global network of chemical companies that delices standards for supply chains | https://tfs-initiative.com/ |
| | Banking/ Finance | *UN Global Compact Sustainable Finance* is bringing together a multi-sector group of corporate finance leaders to develop innovative strategies to | https://www.unglobalcompact.org/sdgs/sustainablefinance |

(continued)

| Category | Sub-category | About | Access |
|---|---|---|---|
| | | mobilize finance towards sustainable development | |
| | Electronics | *The Responsible Business Alliance (RBA)* is comprised of electronics, retail, auto, and toy companies committed to supporting the rights and well-being of workers and communities worldwide affected by the global supply chain | http://www.responsiblebusiness.org/ |
| | Mining/Raw materials | *Responsible minerals' initiative (RMI)* serves as an umbrella organization for the voice of progressive industry to support responsible mineral sourcing broadly and convenes stakeholders to continually shape dialog and practices | http://www.responsiblemineralsinitiative.org/ |
| | Forestry | *Sustainable Forestry Initiative*, an independent, non-profit organization, that collaborate with their network to provide solutions to local and global sustainability challenges | https://www.forests.org/ |
| | Agriculture/Food | *Fair Food Program*, a partnership among farmers, farmworkers, and retail companies that ensure humane wages and working conditions along the value chain | https://www.fairfoodprogram.org/ |
| | Agriculture | *SAI platform*, a non-profit network that aims at advancing sustainable agricultural practices | https://saiplatform.org/ |
| | Food | *International Cocoa initiative* promotes child-protection in cocoa-growing communities | https://cocoainitiative.org |
| | Confectionary/Beauty | *Roundtable on Sustainable Palm Oil*, non-profit that unites stakeholders from several sectors of the palm oil industry | https://rspo.org/ |
| | Rail | *Railsponsible* is an industry initiative focused on sustainable procurement, with the aim to continuously improve sustainability practices throughout the railway industry supply chain | https://www.railsponsible.org/ |

(continued)

| Category | Sub-category | About | Access |
|---|---|---|---|
| | Transport/ Logistics | *IRU* is the world road transport organization. It represents the entire industry, bus, coach, truck, and taxi, and drive the sustainable mobility of people and goods across the planet | https://www.iru.org/ |
| | Oil and gas | *IPIECA* is the global oil and gas industry association for advancing environmental and social performance | https://www.ipieca.org/ |
| | Building/ Construction | *Sustainable Building Alliance's* goal is to develop common metrics that can be used to monitor and compare internationally ecological behavior and sustainable performance through the different voluntary rating schemes. | https://www.buildingrating.org/ |
| Over-arching initiatives | Textiles | The *Partnership for Sustainable Textiles* was founded to improve the social and environmental conditions within global textile production | https://www.textilbuendnis.com/en/ |
| | Leather | Responsible Leather Round Table | https://responsibleleather.org/rlrt/ |
| | Labor conditions | *Fair Labor Association* brings together universities, civil society organizations (CSOs), and companies to find sustainable solutions to systemic labor issues | https://www.fairlabor.org/ |
| | Circular economy and resource efficiency | *Global Alliance on Circular Economy and Resource Efficiency (GACERE)* is one of the deliverables of the EU's Circular Economy Action Plan adopted by the European Commission in March 2020 as part of the European Green Deal | https://ec.europa.eu/environment/international_issues/circular_economy_global_en.htm |
| | Germany Cross-industry | *Econchain*, cross-industry initiative for sustainable practices | https://www.econchain.de/de/ |
| | Bangladesh | The Acord on Fire and Building Safety in Bangladesh | https://bangladeshaccord.org/ |

(continued)

| Category | Sub-category | About | Access |
|---|---|---|---|
| | | One Planet Network | https://www.oneplanetnetwork.org/initiatives |
| Politically driven initiatives | | *Supply Chain Law* ("Lieferkettengesetz"), see the deep-dive article on the Supply Chain Law for more information | https://lieferkettengesetz.de/ |

| Networking platform | About | Access |
|---|---|---|
| SUSTAINX | The supporting ecosystem provides a gateway for companies to build on and guides you from strategy definition towards finding ideas and translating them into action | www.sustainx.de |
| The Sustainable Procurement Pledge (SPP) | The Sustainable Procurement Pledge (SPP) is a non-profit organization that started in 2019 as a movement on LinkedIn. SPP aims at democratizing knowledge about sustainability for procurement professionals | www.spp.earth |

The manufacturer's authorised representative in the EU is Springer Nature Customer Service Centre GmbH, Europaplatz 3, 69115 Heidelberg, Germany. If you have any concerns regarding our products, please contact ProductSafety@springernature.com

Printed and bound by CPI Group (UK) Ltd, Croydon, CR0 4YY
30/09/2026
02225497-0004